Ceramic Materials: Science and Engineering

Ceramic Materials: Science and Engineering

Edited by
Iker Morris

Larsen & Keller
www.larsen-keller.com

Ceramic Materials: Science and Engineering
Edited by Iker Morris
ISBN: 978-1-63549-062-6 (Hardback)

≡ Larsen & Keller

Published by Larsen and Keller Education,
5 Penn Plaza,
19th Floor,
New York, NY 10001, USA

Cataloging-in-Publication Data

Ceramic materials : science and engineering / edited by Iker Morris.
     p. cm.
Includes bibliographical references and index.
ISBN 978-1-63549-062-6
1. Ceramic materials. 2. Ceramic engineering. 3. Ceramics.
4. Ceramic industries. I. Morris, Iker.
TA455.C43 C74 2017
620.14--dc23

The publisher's policy is to use permanent paper from mills that operate a sustainable forestry policy. Furthermore, the publisher ensures that the text paper and cover boards used have met acceptable environmental accreditation standards.

Printed and bound in the United States of America.

For more information regarding Larsen and Keller Education and its products, please visit the publisher's website www.larsen-keller.com

Table of Contents

Preface   **VII**

Chapter 1 **Introduction to Ceramic and Ceramic Engineering**   **1**
- Ceramic   1
- Ceramic Engineering   14
- Ceramography   29

Chapter 2 **Techniques of Ceramic Forming**   **35**
- Ceramic Forming Techniques   35
- Freeze-casting   37
- Freeze Gelation   46
- Sintering   48

Chapter 3 **Innovations in Ceramics**   **63**
- Zirconium Dioxide   63
- Silicon Nitride   69
- Nanoceramic   75
- Coade Stone   77
- Ceramic Matrix Composite   83
- Transparent Ceramics   97

Chapter 4 **Strength of Ceramics: A Comprehensive Study**   **112**
- Grain Boundary Strengthening   112
- Strengthening Mechanisms of Materials   116
- Solid Solution Strengthening   123

Chapter 5 **Properties of Ceramics**   **126**
- High-temperature Superconductivity   126
- Piezoelectricity   139
- Compressive Strength   155
- Stiffness   158

Chapter 6 **Various Ceramic Materials**   **166**
- Glass-Ceramic   166
- Glass-to-Metal Seal   172
- Terracotta   180
- Architectural Terracotta   186

Chapter 7 **Integrating Materials Science in Ceramic Engineering**   **191**
- Materials Science   191
- Composite Material   205
- Nanomaterials   217

Chapter 8 **Diverse Applications of Ceramic Engineering**   **225**
           • Bulletproof Vest   225
           • Implant (Medicine)   250
           • Artificial Bone   254
           • Insulator (Electricity)   255
           • Airframe   266

           Permissions

           Index

Preface

This book provides comprehensive insights into the field of ceramic materials. It discusses in detail the different concepts and theories related to this subject. Ceramic engineering is the field which deals with the design and manufacture of objects from non-metallic and inorganic materials. This field of study is especially used in other engineering subjects like mechanical engineering, materials engineering, chemical engineering, etc. This text is a valuable compilation of topics, ranging from the basic to the most complex principles and practices in the field of ceramic science and engineering. Those in search of information to further their knowledge will be greatly assisted by this textbook.

Given below is the chapter wise description of the book:

Chapter 1- A ceramic is a solid material that is made up of metal, non-metal or metalloid atoms. The study of creating objects from inorganic or non-metallic materials is termed as ceramic engineering. This chapter will provide an integrated understanding of ceramic and ceramic engineering.

Chapter 2- Freeze gelation helps in the processing of a ceramic object to be fabricated in complex shapes. Alternatively, the techniques used in ceramic forming are freeze-casting and sintering. The aspects elucidated in this chapter are of vital importance and provides a better understanding of ceramic farming.

Chapter 3- This section focuses on the innovations in ceramics. Some of these innovations are zirconium dioxide, boron carbide, silicon nitride, nanoceramic and coade stones. The basic purpose of zirconium dioxide is the production of ceramics whereas silicon nitride bearings are full ceramic bearings. Silicon nitride ceramics has good shock resistance as compared to other ceramics. This section has been carefully written to provide an in-depth understanding of the use of ceramics.

Chapter 4- Grain-boundary strengthening is a technique of strengthening materials by basically changing their average grain size. The alternative methods used for strengthening ceramics are strengthening mechanisms of materials and solid solution strengthening. This chapter is an overview of the subject matter incorporating all the major aspects of strength of ceramics.

Chapter 5- The properties of ceramics are high-temperature superconductivity, piezoelectricity, compressive strength, indentation hardness and stiffness. High-temperature superconductivity is the manner in which materials behave at high temperatures. The properties elucidated in this section are of vital importance and provides a better understanding of ceramics.

Chapter 6- Glass-ceramics are produced by controlled crystallizations. Glass-ceramics have the advantage of both glass as well as ceramics. Some of the examples of glass-ceramics are bioactive glass, pyroceram, glass-ceramic-to-metal seals and sitall. This chapter helps the readers in understanding the various materials created by ceramics.

Chapter 7- Materials science and engineering involves the process of discovery and designing of new materials. The emphasis in this process is on solids. The topics elucidated in this chapter are composite materials and nanomaterials. These topics help in broadening the existing the knowledge on ceramic engineering.

Chapter 8- The diverse applications of ceramic engineering are bulletproof vest, medical implant, artificial bone, insulator and airframe. Ceramic engineering is an emerging field of study; the following chapter will not only provide an overview, it will also delve deep into the applications related to it.

At the end, I would like to thank all those who dedicated their time and efforts for the successful completion of this book. I also wish to convey my gratitude towards my friends and family who supported me at every step.

Editor

Introduction to Ceramic and Ceramic Engineering

A ceramic is a solid material that is made up of metal, non-metal or metalloid atoms. The study of creating objects from inorganic or non-metallic materials is termed as ceramic engineering. This chapter will provide an integrated understanding of ceramic and ceramic engineering.

Ceramic

A ceramic is an inorganic, nonmetallic solid material comprising metal, nonmetal or metalloid atoms primarily held in ionic and covalent bonds. The crystallinity of ceramic materials ranges from highly oriented to semi-crystalline, and often completely amorphous (e.g., glasses). Varying crystallinity and electron consumption in the ionic and covalent bonds cause most ceramic materials to be good thermal and electrical insulators (extensively researched in ceramic engineering). With such a large range of possible options for the composition/structure of a ceramic (e.g. nearly all of the elements, nearly all types of bonding, and all levels of crystallinity), the breadth of the subject is vast, and identifiable attributes (e.g. hardness, toughness, electrical conductivity, etc.) are hard to specify for the group as a whole. General properties such as high melting temperature, high hardness, poor conductivity, high moduli of elasticity, chemical resistance and low ductility are the norm, with known exceptions to each of these rules (e.g. piezoelectric ceramics, glass transition temperature, superconductive ceramics, etc.). Many composites, such as fiberglass and carbon fiber, while containing ceramic materials, are not considered to be part of the ceramic family.

A Ming Dynasty porcelain vase dated to 1403–1424

The word "ceramic" may be used as an adjective to describe a material, product or process, or it may be used as a noun, either singular, or, more commonly, as the plural noun "ceramics".

A selection of silicon nitride components.

Spherical Hanging Ornament, 1575-1585, Ottoman Period. Brooklyn Museum.

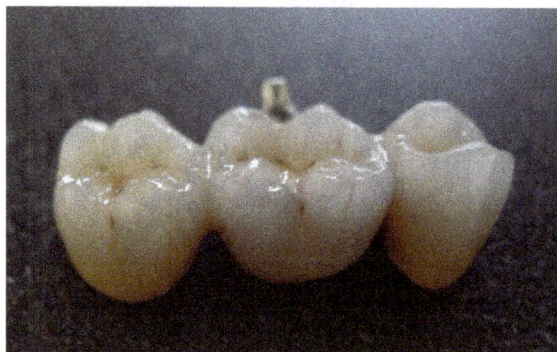

Fixed partial porcelain denture, or "bridge"

The earliest ceramics made by humans were pottery objects, including 27,000-year-old figurines, made from clay, either by itself or mixed with other materials like silica, hardened, sintered, in fire. Later ceramics were glazed and fired to create smooth, colored surfaces, decreasing porosity through the use of glassy, amorphous ceramic coatings on top of the crystalline ceramic substrates. Ceramics now include domestic, industrial and building products, as well as a wide range of ceramic art. In the 20th century, new ceramic materials were developed for use in advanced ceramic engineering, such as in semiconductors.

Types of Ceramic Material

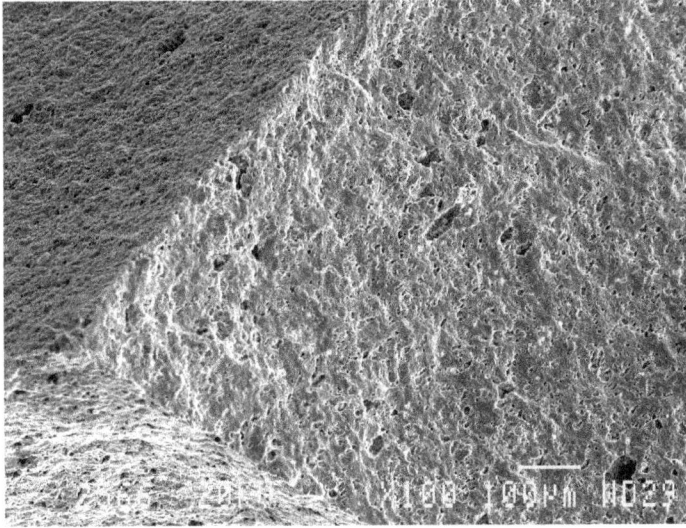

A low magnification SEM micrograph of an advanced ceramic material. The properties of ceramics make fracturing an important inspection method.

A ceramic material is an inorganic, non-metallic, often crystalline oxide, nitride or carbide material. Some elements, such as carbon or silicon, may be considered ceramics. Ceramic materials are brittle, hard, strong in compression, weak in shearing and tension. They withstand chemical erosion that occurs in other materials subjected to acidic or caustic environments. Ceramics generally can withstand very high temperatures, such as temperatures that range from 1,000 °C to 1,600 °C (1,800 °F to 3,000 °F). Glass is often not considered a ceramic because of its amorphous (non-crystalline) character. However, glassmaking involves several steps of the ceramic process and its mechanical properties are similar to ceramic materials.

Traditional ceramic raw materials include clay minerals such as kaolinite, whereas more recent materials include aluminium oxide, more commonly known as alumina. The modern ceramic materials, which are classified as advanced ceramics, include silicon carbide and tungsten carbide. Both are valued for their abrasion resistance, and hence find use in applications such as the wear plates of crushing equipment in mining operations. Advanced ceramics are also used in the medicine, electrical, electronics industries and body armor.

Crystalline Ceramics

Crystalline ceramic materials are not amenable to a great range of processing. Methods for dealing with them tend to fall into one of two categories – either make the ceramic in the desired shape, by reaction *in situ*, or by "forming" powders into the desired shape, and then sintering to form a solid body. Ceramic forming techniques include shaping by hand (sometimes including a rotation process called "throwing"), slip casting, tape casting (used for making very thin ceramic capacitors, e.g.), injection molding, dry pressing, and other variations. Details of these processes are described in the two books listed below. A few methods use a hybrid between the two approaches.

Noncrystalline Ceramics

Noncrystalline ceramics, being glass, tend to be formed from melts. The glass is shaped when either fully molten, by casting, or when in a state of toffee-like viscosity, by methods such as blowing into a mold. If later heat treatments cause this glass to become partly crystalline, the resulting material is known as a glass-ceramic, widely used as cook-top and also as a glass composite material for nuclear waste disposal.

Properties of Ceramics

The physical properties of any ceramic substance are a direct result of its crystalline structure and chemical composition. Solid state chemistry reveals the fundamental connection between microstructure and properties such as localized density variations, grain size distribution, type of porosity and second-phase content, which can all be correlated with ceramic properties such as mechanical strength σ by the Hall-Petch equation, hardness, toughness, dielectric constant, and the optical properties exhibited by transparent materials.

Physical properties of chemical compounds which provide evidence of chemical composition include odor, colour, volume, density (mass / volume), melting point, boiling point, heat capacity, physical form at room temperature (solid, liquid or gas), hardness, porosity, and index of refraction.

Ceramography is the art and science of preparation, examination and evaluation of ceramic microstructures. Evaluation and characterization of ceramic microstructures is often implemented on similar spatial scales to that used commonly in the emerging field of nanotechnology: from tens of angstroms (A) to tens of micrometers (μm). This is typically somewhere between the minimum wavelength of visible light and the resolution limit of the naked eye.

The microstructure includes most grains, secondary phases, grain boundaries, pores, micro-cracks, structural defects and hardness microindentions. Most bulk mechanical, optical, thermal, electrical and magnetic properties are significantly affected by the observed microstructure. The fabrication method and process conditions are generally indicated by the microstructure. The root cause of many ceramic failures is evident in the cleaved and polished microstructure. Physical properties which constitute the field of materials science and engineering include the following:

Mechanical Properties

Mechanical properties are important in structural and building materials as well as textile fabrics. They include the many properties used to describe the strength of materials such as: elasticity / plasticity, tensile strength, compressive strength, shear strength, fracture toughness & ductility (low in brittle materials), and indentation hardness.

Cutting disks made of silicon carbide

The Porsche Carrera GT's carbon-ceramic (silicon carbide) disc brake

In modern materials science, fracture mechanics is an important tool in improving the mechanical performance of materials and components. It applies the physics of stress and strain, in particular the theories of elasticity and plasticity, to the microscopic crystallographic defects found in real materials in order to predict the macroscopic mechanical failure of bodies. Fractography is widely used with fracture mechanics to understand the causes of failures and also verify the theoretical failure predictions with real life failures.

Ceramic materials are usually ionic or covalent bonded materials, and can be crystalline or amorphous. A material held together by either type of bond will tend to fracture before any plastic deformation takes place, which results in poor toughness in these materials. Additionally, because these materials tend to be porous, the pores and other microscopic imperfections act as stress concentrators, decreasing the toughness further, and reducing the tensile strength. These combine to give catastrophic failures, as opposed to the normally much more gentle failure modes of metals.

These materials do show plastic deformation. However, due to the rigid structure of the crystalline materials, there are very few available slip systems for dislocations to move, and so they deform very slowly. With the non-crystalline (glassy) materials, viscous flow is the dominant source of plastic deformation, and is also very slow. It is therefore neglected in many applications of ceramic materials.

To overcome the brittle behaviour, ceramic material development has introduced the class of ceramic matrix composite materials, in which ceramic fibers are embedded and with specific coatings are forming fiber bridges across any crack. This mechanism substantially increases the fracture toughness of such ceramics. The ceramic disc brakes are, for example using a ceramic matrix composite material manufactured with a specific process.

Electrical Properties

Semiconductors

Some ceramics are semiconductors. Most of these are transition metal oxides that are II-VI semiconductors, such as zinc oxide.

While there are prospects of mass-producing blue LEDs from zinc oxide, ceramicists are most interested in the electrical properties that show grain boundary effects.

One of the most widely used of these is the varistor. These are devices that exhibit the property that resistance drops sharply at a certain threshold voltage. Once the voltage across the device reaches the threshold, there is a breakdown of the electrical structure in the vicinity of the grain boundaries, which results in its electrical resistance dropping from several megohms down to a few hundred ohms. The major advantage of these is that they can dissipate a lot of energy, and they self-reset – after the voltage across the device drops below the threshold, its resistance returns to being high.

This makes them ideal for surge-protection applications; as there is control over the threshold voltage and energy tolerance, they find use in all sorts of applications. The best demonstration of their ability can be found in electrical substations, where they are employed to protect the infrastructure from lightning strikes. They have rapid response, are low maintenance, and do not appreciably degrade from use, making them virtually ideal devices for this application.

Semiconducting ceramics are also employed as gas sensors. When various gases are passed over a polycrystalline ceramic, its electrical resistance changes. With tuning to the possible gas mixtures, very inexpensive devices can be produced.

Superconductivity

The Meissner effect demonstrated by levitating a magnet above a cuprate superconductor, which is cooled by liquid nitrogen

Under some conditions, such as extremely low temperature, some ceramics exhibit high temperature superconductivity. The exact reason for this is not known, but there are two major families of superconducting ceramics.

Ferroelectricity and Supersets

Piezoelectricity, a link between electrical and mechanical response, is exhibited by a large number of ceramic materials, including the quartz used to measure time in watches and other electronics. Such devices use both properties of piezoelectrics, using electricity to produce a mechanical motion (powering the device) and then using this mechanical motion to produce electricity (generating a signal). The unit of time measured is the natural interval required for electricity to be converted into mechanical energy and back again.

The piezoelectric effect is generally stronger in materials that also exhibit pyroelectricity, and all pyroelectric materials are also piezoelectric. These materials can be used to inter convert between thermal, mechanical, or electrical energy; for instance, after synthesis in a furnace, a pyroelectric crystal allowed to cool under no applied stress generally builds up a static charge of thousands of volts. Such materials are used in motion sensors, where the tiny rise in temperature from a warm body entering the room is enough to produce a measurable voltage in the crystal.

In turn, pyroelectricity is seen most strongly in materials which also display the ferroelectric effect, in which a stable electric dipole can be oriented or reversed by applying an electrostatic field. Pyroelectricity is also a necessary consequence of ferroelectricity. This can be used to store information in ferroelectric capacitors, elements of ferroelectric RAM.

The most common such materials are lead zirconate titanate and barium titanate. Aside from the uses mentioned above, their strong piezoelectric response is exploited in the design of high-frequency loudspeakers, transducers for sonar, and actuators for atomic force and scanning tunneling microscopes.

Positive Thermal Coefficient

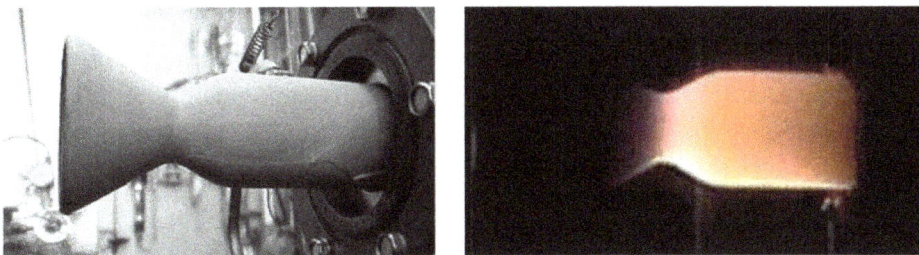

Silicon nitride rocket thruster. Left: Mounted in test stand. Right: Being tested with H_2/O_2 propellants

Increases in temperature can cause grain boundaries to suddenly become insulating in some semiconducting ceramic materials, mostly mixtures of heavy metal titanates. The critical transition temperature can be adjusted over a wide range by variations in chemistry. In such materials, current will pass through the material until joule heating brings it to the transition temperature, at which point the circuit will be broken and current flow will cease. Such ceramics are used as self-controlled heating elements in, for example, the rear-window defrost circuits of automobiles.

At the transition temperature, the material's dielectric response becomes theoretically infinite. While a lack of temperature control would rule out any practical use of the material near its critical temperature, the dielectric effect remains exceptionally strong even at much higher temperatures. Titanates with critical temperatures far below room temperature have become synonymous with "ceramic" in the context of ceramic capacitors for just this reason.

Optical Properties

Cermax xenon arc lamp with synthetic sapphire output window

Optically transparent materials focus on the response of a material to incoming lightwaves of a range of wavelengths. Frequency selective optical filters can be utilized to alter or enhance the brightness and contrast of a digital image. Guided lightwave transmission via frequency selective waveguides involves the emerging field of fiber optics and the ability of certain glassy compositions as a transmission medium for a range of frequencies simultaneously (multi-mode optical fiber) with little or no interference between competing wavelengths or frequencies. This resonant mode of energy and data transmission via electromagnetic (light) wave propagation, though low powered, is virtually lossless. Optical waveguides are used as components in Integrated optical circuits (e.g. light-emitting diodes, LEDs) or as the transmission medium in local and long haul optical communication systems. Also of value to the emerging materials scientist is the sensitivity of materials to radiation in the thermal infrared (IR) portion of the electromagnetic spectrum. This heat-seeking ability is responsible for such diverse optical phenomena as Night-vision and IR luminescence.

Thus, there is an increasing need in the military sector for high-strength, robust materials which have the capability to transmit light (electromagnetic waves) in the visible (0.4 – 0.7 micrometers) and mid-infrared (1 – 5 micrometers) regions of the spectrum. These materials are needed for applications requiring transparent armor, including next-generation high-speed missiles and pods, as well as protection against improvised explosive devices (IED).

In the 1960s, scientists at General Electric (GE) discovered that under the right manufacturing conditions, some ceramics, especially aluminium oxide (alumina), could be made translucent. These translucent materials were transparent enough to be used for containing the electrical plasma generated in high-pressure sodium street lamps. During the past two decades, additional types of transparent ceramics have been developed for applications such as nose cones for heat-seeking missiles, windows for fighter aircraft, and scintillation counters for computed tomography scanners.

In the early 1970s, Thomas Soules pioneered computer modeling of light transmission through translucent ceramic alumina. His model showed that microscopic pores in ceramic, mainly trapped at the junctions of microcrystalline grains, caused light to scatter and prevented true transparency. The volume fraction of these microscopic pores had to be less than 1% for high-quality optical transmission.

This is basically a particle size effect. Opacity results from the incoherent scattering of light at surfaces and interfaces. In addition to pores, most of the interfaces in a typical metal or ceramic object are in the form of grain boundaries which separate tiny regions of crystalline order. When the size of the scattering center (or grain boundary) is reduced below the size of the wavelength of the light being scattered, the scattering no longer occurs to any significant extent.

In the formation of polycrystalline materials (metals and ceramics) the size of the crystalline grains is determined largely by the size of the crystalline particles present in the raw material during formation (or pressing) of the object. Moreover, the size of the grain boundaries scales directly with particle size. Thus a reduction of the original particle size below the wavelength of visible light (~ 0.5 micrometers for shortwave violet) eliminates any light scattering, resulting in a transparent material.

Recently, Japanese scientists have developed techniques to produce ceramic parts that rival the transparency of traditional crystals (grown from a single seed) and exceed the fracture toughness of a single crystal. In particular, scientists at the Japanese firm Konoshima Ltd., a producer of ceramic construction materials and industrial chemicals, have been looking for markets for their transparent ceramics.

Livermore researchers realized that these ceramics might greatly benefit high-powered lasers used in the National Ignition Facility (NIF) Programs Directorate. In particular, a Livermore research team began to acquire advanced transparent ceramics from Konoshima to determine if they could meet the optical requirements needed for Livermore's Solid-State Heat Capacity Laser (SSHCL). Livermore researchers have also been testing applications of these materials for applications such as advanced drivers for laser-driven fusion power plants.

Examples

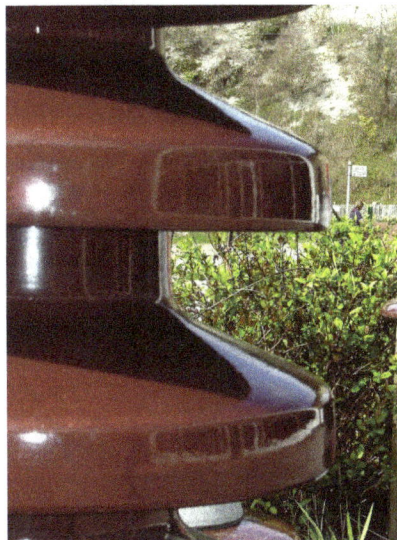

Porcelain high-voltage insulator

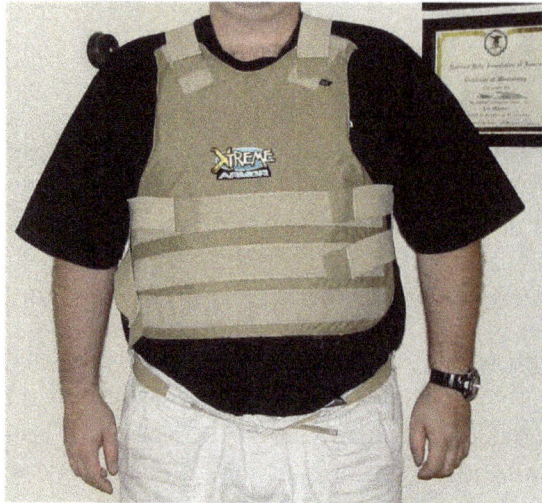

Silicon carbide is used for inner plates of ballistic vests

Until the 1950s, the most important ceramic materials were (1) pottery, bricks and tiles, (2) cements and (3) glass. A composite material of ceramic and metal is known as cermet.

Other ceramic materials, generally requiring greater purity in their make-up than those above, include forms of several chemical compounds, including:

- Barium titanate (often mixed with strontium titanate) displays ferroelectricity, meaning that its mechanical, electrical, and thermal responses are coupled to one another and also history-dependent. It is widely used in electromechanical transducers, ceramic capacitors, and data storage elements. Grain boundary conditions can create PTC effects in heating elements.

- Bismuth strontium calcium copper oxide, a high-temperature superconductor

- Boron oxide is used in body armor.

- Boron nitride is structurally isoelectronic to carbon and takes on similar physical forms: a graphite-like one used as a lubricant, and a diamond-like one used as an abrasive.

- Earthenware used for domestic ware such as plates and mugs.

- Ferrite is used in the magnetic cores of electrical transformers and magnetic core memory.

- Lead zirconate titanate (PZT) was developed at the United States National Bureau of Standards in 1954. PZT is used as an ultrasonic transducer, as its piezoelectric properties greatly exceed those of Rochelle salt.

- Magnesium diboride (MgB_2) is an unconventional superconductor.

- Porcelain is used for a wide range of household and industrial products.

- Sialon (Silicon Aluminium Oxynitride) has high strength; resistance to thermal shock, chemical and wear resistance, and low density. These ceramics are used in non-ferrous molten metal handling, weld pins and the chemical industry.

- Silicon carbide (SiC) is used as a susceptor in microwave furnaces, a commonly used abrasive, and as a refractory material.

- Silicon nitride (Si_3N_4) is used as an abrasive powder.

- Steatite (magnesium silicates) is used as an electrical insulator.

- Titanium carbide Used in space shuttle re-entry shields and scratchproof watches.

- Uranium oxide (UO_2), used as fuel in nuclear reactors.

- Yttrium barium copper oxide ($YBa_2Cu_3O_{7-x}$), another high temperature superconductor.

- Zinc oxide (ZnO), which is a semiconductor, and used in the construction of varistors.

- Zirconium dioxide (zirconia), which in pure form undergoes many phase changes between room temperature and practical sintering temperatures, can be chemically "stabilized" in several different forms. Its high oxygen ion conductivity recommends it for use in fuel cells and automotive oxygen sensors. In another variant, metastable structures can impart transformation toughening for mechanical applications; most ceramic knife blades are made of this material.

- Partially stabilised zirconia (PSZ) is much less brittle than other ceramics and is used for metal forming tools, valves and liners, abrasive slurries, kitchen knives and bearings subject to severe abrasion.

Kitchen knife with a ceramic blade

Ceramic Products

By Usage

For convenience, ceramic products are usually divided into four main types; these are shown below with some examples:

- Structural, including bricks, pipes, floor and roof tiles

- Refractories, such as kiln linings, gas fire radiants, steel and glass making crucibles

- Whitewares, including tableware, cookware, wall tiles, pottery products and sanitary ware

- Technical, also known as engineering, advanced, special, and fine ceramics. Such items include:

  o   gas burner nozzles

- o ballistic protection

- o nuclear fuel uranium oxide pellets

- o biomedical implants

- o coatings of jet engine turbine blades

- o ceramic disk brake

- o missile nose cones

- o bearing (mechanical)

- o tiles used in the Space Shuttle program

Ceramics Made with Clay

Frequently, the raw materials of modern ceramics do not include clays. Those that do are classified as follows:

- Earthenware, fired at lower temperatures than other types

- Stoneware, vitreous or semi-vitreous

- Porcelain, which contains a high content of kaolin

- Bone china

Classification of Technical Ceramics

Technical ceramics can also be classified into three distinct material categories:

- Oxides: alumina, beryllia, ceria, zirconia

- Nonoxides: carbide, boride, nitride, silicide

- Composite materials: particulate reinforced, fiber reinforced, combinations of oxides and nonoxides.

Each one of these classes can develop unique material properties because ceramics tend to be crystalline.

Applications

- Knife blades: the blade of a ceramic knife will stay sharp for much longer than that of a steel knife, although it is more brittle and can snap from a fall onto a hard surface.

- Carbon-ceramic brake disks for vehicles are resistant to brake fade at high temperatures.

- Advanced composite ceramic and metal matrices have been designed for most modern armoured fighting vehicles because they offer superior penetrating resistance against shaped charges (such as HEAT rounds) and kinetic energy penetrators.

- Ceramics such as alumina and boron carbide have been used in ballistic armored vests to repel large-caliber rifle fire. Such plates are known commonly as small arms protective inserts, or SAPIs. Similar material is used to protect the cockpits of some military airplanes, because of the low weight of the material.

- Ceramics can be used in place of steel for ball bearings. Their higher hardness means they are much less susceptible to wear and typically last for triple the lifetime of a steel part. They also deform less under load, meaning they have less contact with the bearing retainer walls and can roll faster. In very high speed applications, heat from friction during rolling can cause problems for metal bearings, which are reduced by the use of ceramics. Ceramics are also more chemically resistant and can be used in wet environments where steel bearings would rust. In some cases, their electricity-insulating properties may also be valuable in bearings. Two drawbacks to ceramic bearings are a significantly higher cost and susceptibility to damage under shock loads.

- In the early 1980s, Toyota researched production of an adiabatic engine using ceramic components in the hot gas area. The ceramics would have allowed temperatures of over 3000 °F (1650 °C). The expected advantages would have been lighter materials and a smaller cooling system (or no need for one at all), leading to a major weight reduction. The expected increase of fuel efficiency of the engine (caused by the higher temperature, as shown by Carnot's theorem) could not be verified experimentally; it was found that the heat transfer on the hot ceramic cylinder walls was higher than the transfer to a cooler metal wall as the cooler gas film on the metal surface works as a thermal insulator. Thus, despite all of these desirable properties, such engines have not succeeded in production because of costs for the ceramic components and the limited advantages. (Small imperfections in the ceramic material with its low fracture toughness lead to cracks, which can lead to potentially dangerous equipment failure.) Such engines are possible in laboratory settings, but mass production is not feasible with current technology.

- Work is being done in developing ceramic parts for gas turbine engines. Currently, even blades made of advanced metal alloys used in the engines' hot section require cooling and careful limiting of operating temperatures. Turbine engines made with ceramics could operate more efficiently, giving aircraft greater range and payload for a set amount of fuel.

- Recent advances have been made in ceramics which include bioceramics, such as dental implants and synthetic bones. Hydroxyapatite, the natural mineral component of bone, has been made synthetically from a number of biological and chemical sources and can be formed into ceramic materials. Orthopedic implants coated with these materials bond readily to bone and other tissues in the body without rejection or inflammatory reactions so are of great interest for gene delivery and tissue engineering scaffolds. Most hydroxyapatite ceramics are very porous and lack mechanical strength, and are used to coat metal orthopedic devices to aid in forming a bond to bone or as bone fillers. They are also used as fillers for orthopedic plastic screws to aid in reducing the inflammation and increase absorption of these plastic materials. Work is being done to make strong, fully dense nanocrystalline hydroxyapatite ceramic materials for orthopedic weight bearing devices, replacing foreign metal and plastic orthopedic materials with a synthetic, but naturally occurring, bone mineral. Ultimately, these ceramic materials may be used as bone replacements or with the incorporation of protein collagens, synthetic bones.

- High-tech ceramic is used in watchmaking for producing watch cases. The material is valued by watchmakers for its light weight, scratch resistance, durability and smooth touch. IWC is one of the brands that initiated the use of ceramic in watchmaking.

Ceramics in Archaeology

Ceramic artifacts have an important role in archaeology for understanding the culture, technology and behavior of peoples of the past. They are among the most common artifacts to be found at an archaeological site, generally in the form of small fragments of broken pottery called sherds. Processing of collected sherds can be consistent with two main types of analysis: technical and traditional.

Traditional analysis involves sorting ceramic artifacts, sherds and larger fragments into specific types based on style, composition, manufacturing and morphology. By creating these typologies it is possible to distinguish between different cultural styles, the purpose of the ceramic and technological state of the people among other conclusions. In addition, by looking at stylistic changes of ceramics over time is it possible to separate (seriate) the ceramics into distinct diagnostic groups (assemblages). A comparison of ceramic artifacts with known dated assemblages allows for a chronological assignment of these pieces.

The technical approach to ceramic analysis involves a finer examination of the composition of ceramic artifacts and sherds to determine the source of the material and through this the possible manufacturing site. Key criteria are the composition of the clay and the temper used in the manufacture of the article under study: temper is a material added to the clay during the initial production stage, and it is used to aid the subsequent drying process. Types of temper include shell pieces, granite fragments and ground sherd pieces called 'grog'. Temper is usually identified by microscopic examination of the temper material. Clay identification is determined by a process of refiring the ceramic, and assigning a color to it using Munsell Soil Color notation. By estimating both the clay and temper compositions, and locating a region where both are known to occur, an assignment of the material source can be made. From the source assignment of the artifact further investigations can be made into the site of manufacture.

Ceramic Engineering

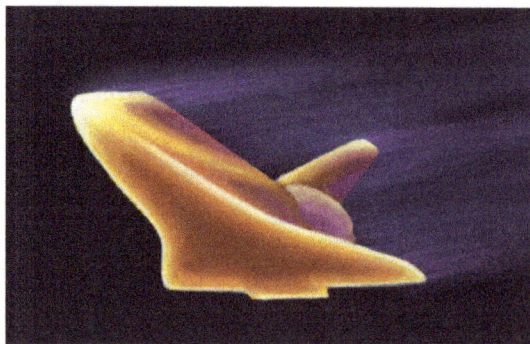

Simulation of the outside of the Space Shuttle as it heats up to over 1,500 °C (2,730 °F) during re-entry into the Earth's atmosphere

Ceramic bread knife

Ceramic materials may have a crystalline or partly crystalline structure, with long-range order on atomic scale. Glass ceramics may have an amorphous or glassy structure, with limited or short-range atomic order. They are either formed from a molten mass that solidifies on cooling, formed and matured by the action of heat, or chemically synthesized at low temperatures using, for example, hydrothermal or sol-gel synthesis.

The special character of ceramic materials gives rise to many applications in materials engineering, electrical engineering, chemical engineering and mechanical engineering. As ceramics are heat resistant, they can be used for many tasks for which materials like metal and polymers are unsuitable. Ceramic materials are used in a wide range of industries, including mining, aerospace, medicine, refinery, food and chemical industries, packaging science, electronics, industrial and transmission electricity, and guided lightwave transmission.

History

The plural "ceramics" may be used to refer the making of things out of ceramic materials. Ceramic engineering, like many sciences, evolved from a different discipline by today's standards. Materials science engineering is grouped with ceramics engineering to this day.

Abraham Darby first used coke in 1709 in Shropshire, England, to improve the yield of a smelting process. Coke is now widely used to produce carbide ceramics. Potter Josiah Wedgwood opened the first modern ceramics factory in Stoke-on-Trent, England, in 1759. Austrian chemist Carl Josef Bayer, working for the textile industry in Russia, developed a process to separate alumina from bauxite ore in 1888. The Bayer process is still used to purify alumina for the ceramic and aluminium industries. Brothers Pierre and Jacques Curie discovered piezoelectricity in Rochelle salt circa 1880. Piezoelectricity is one of the key properties of electroceramics.

E.G. Acheson heated a mixture of coke and clay in 1893, and invented carborundum, or synthetic silicon carbide. Henri Moissan also synthesized SiC and tungsten carbide in his electric arc furnace in Paris about the same time as Acheson. Karl Schröter used liquid-phase sintering to bond or "cement" Moissan's tungsten carbide particles with cobalt in 1923 in Germany. Cemented (metal-bonded) carbide edges greatly increase the durability of hardened steel cutting tools. W.H. Nernst developed cubic-stabilized zirconia in the 1920s in Berlin. This material is used as an oxygen sensor in exhaust systems. The main limitation on the use of ceramics in engineering is brittleness.

Military

Soldiers pictured during the 2003 Iraq War seen through IR transparent Night Vision Goggles

The military requirements of World War II encouraged developments, which created a need for high-performance materials and helped speed the development of ceramic science and engineering. Throughout the 1960s and 1970s, new types of ceramics were developed in response to advances in atomic energy, electronics, communications, and space travel. The discovery of ceramic superconductors in 1986 has spurred intense research to develop superconducting ceramic parts for electronic devices, electric motors, and transportation equipment.

There is an increasing need in the military sector for high-strength, robust materials which have the capability to transmit light around the visible (0.4–0.7 micrometers) and mid-infrared (1–5 micrometers) regions of the spectrum. These materials are needed for applications requiring transparent armour. Transparent armor is a material or system of materials designed to be optically transparent, yet protect from fragmentation or ballistic impacts. The primary requirement for a transparent armour system is to not only defeat the designated threat but also provide a multi-hit capability with minimized distortion of surrounding areas. Transparent armour windows must also be compatible with night vision equipment. New materials that are thinner, lightweight, and offer better ballistic performance are being sought. Such solid-state components have found widespread use for various applications in the electro-optical field including: optical fibres for guided lightwave transmission, optical switches, laser amplifiers and lenses, hosts for solid-state lasers and optical window materials for gas lasers, and infrared (IR) heat seeking devices for missile guidance systems and IR night vision.

Modern Industry

Now a multibillion-dollar a year industry, ceramic engineering and research has established itself as an important field of science. Applications continue to expand as researchers develop new kinds of ceramics to serve different purposes.

- Zirconium dioxide ceramics are used in the manufacture of knives. The blade of the ceramic knife will stay sharp for much longer than that of a steel knife, although it is more brittle and can be snapped by dropping it on a hard surface.

- Ceramics such as alumina, boron carbide and silicon carbide have been used in bulletproof vests to repel small arms rifle fire. Such plates are known commonly as trauma plates. Similar material is used to protect cockpits of some military aircraft, because of the low weight of the material.

- Silicon nitride parts are used in ceramic ball bearings. Their higher hardness means that they are much less susceptible to wear and can offer more than triple lifetimes. They also deform less under load meaning they have less contact with the bearing retainer walls and can roll faster. In very high speed applications, heat from friction during rolling can cause problems for metal bearings; problems which are reduced by the use of ceramics. Ceramics are also more chemically resistant and can be used in wet environments where steel bearings would rust. The major drawback to using ceramics is a significantly higher cost. In many cases their electrically insulating properties may also be valuable in bearings.

- In the early 1980s, Toyota researched production of an adiabatic ceramic engine which can run at a temperature of over 6000 °F (3300 °C). Ceramic engines do not require a cooling system and hence allow a major weight reduction and therefore greater fuel efficiency. Fuel efficiency of the engine is also higher at high temperature, as shown by Carnot's theorem. In a conventional metallic engine, much of the energy released from the fuel must be dissipated as waste heat in order to prevent a meltdown of the metallic parts. Despite all of these desirable properties, such engines are not in production because the manufacturing of ceramic parts in the requisite precision and durability is difficult. Imperfection in the ceramic leads to cracks, which can lead to potentially dangerous equipment failure. Such engines are possible in laboratory settings, but mass-production is not feasible with current technology.

- Work is being done in developing ceramic parts for gas turbine engines. Currently, even blades made of advanced metal alloys used in the engines' hot section require cooling and careful limiting of operating temperatures. Turbine engines made with ceramics could operate more efficiently, giving aircraft greater range and payload for a set amount of fuel.

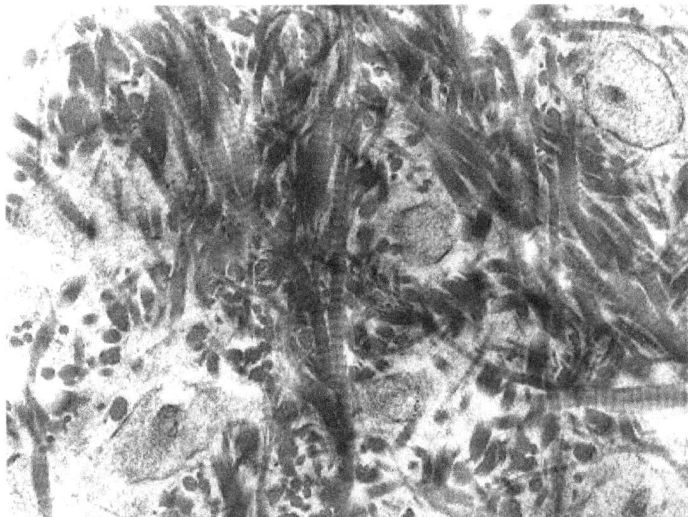

Collagen fibers of woven bone

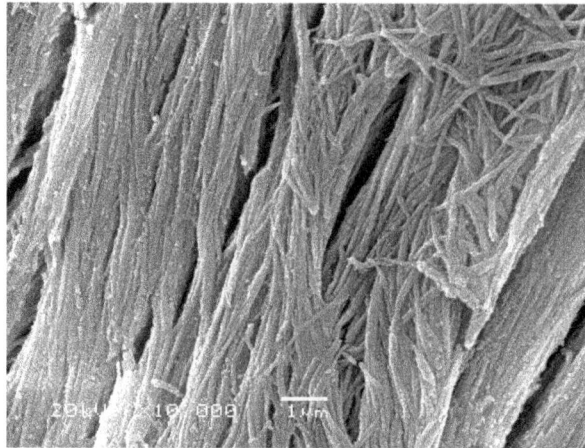

Scanning electron microscopy image of bone

- Recently, there have been advances in ceramics which include bio-ceramics, such as dental implants and synthetic bones. Hydroxyapatite, the natural mineral component of bone, has been made synthetically from a number of biological and chemical sources and can be formed into ceramic materials. Orthopaedic implants made from these materials bond readily to bone and other tissues in the body without rejection or inflammatory reactions. Because of this, they are of great interest for gene delivery and tissue engineering scaffolds. Most hydroxyapatite ceramics are very porous and lack mechanical strength and are used to coat metal orthopaedic devices to aid in forming a bond to bone or as bone fillers. They are also used as fillers for orthopaedic plastic screws to aid in reducing the inflammation and increase absorption of these plastic materials. Work is being done to make strong, fully dense nano crystalline hydroxyapatite ceramic materials for orthopaedic weight bearing devices, replacing foreign metal and plastic orthopaedic materials with a synthetic, but naturally occurring, bone mineral. Ultimately these ceramic materials may be used as bone replacements or with the incorporation of protein collagens, synthetic bones.

- High-tech ceramic is used in watch-making for producing watch cases. The material is valued by watchmakers for its light weight, scratch-resistance, durability and smooth touch. IWC is one of the brands that initiated the use of ceramic in watch-making. The case of the IWC 2007 Top Gun edition of the Pilot's Watch Double chronograph is crafted in high-tech black ceramic.

Glass-ceramics

A high strength glass-ceramic cook-top with negligible thermal expansion.

Glass-ceramic materials share many properties with both glasses and ceramics. Glass-ceramics have an amorphous phase and one or more crystalline phases and are produced by a so-called "controlled crystallization", which is typically avoided in glass manufacturing. Glass-ceramics often contain a crystalline phase which constitutes anywhere from 30% [m/m] to 90% [m/m] of its composition by volume, yielding an array of materials with interesting thermomechanical properties.

In the processing of glass-ceramics, molten glass is cooled down gradually before reheating and annealing. In this heat treatment the glass partly crystallizes. In many cases, so-called 'nucleation agents' are added in order to regulate and control the crystallization process. Because there is usually no pressing and sintering, glass-ceramics do not contain the volume fraction of porosity typically present in sintered ceramics.

The term mainly refers to a mix of lithium and aluminosilicates which yields an array of materials with interesting thermomechanical properties. The most commercially important of these have the distinction of being impervious to thermal shock. Thus, glass-ceramics have become extremely useful for countertop cooking. The negative thermal expansion coefficient (TEC) of the crystalline ceramic phase can be balanced with the positive TEC of the glassy phase. At a certain point (~70% crystalline) the glass-ceramic has a net TEC near zero. This type of glass-ceramic exhibits excellent mechanical properties and can sustain repeated and quick temperature changes up to 1000 °C.

Processing Steps

The traditional ceramic process generally follows this sequence: Milling → Batching → Mixing → Forming → Drying → Firing → Assembly.

Ball mill

- Milling is the process by which materials are reduced from a large size to a smaller size. Milling may involve breaking up cemented material (in which case individual particles retain their shape) or pulverization (which involves grinding the particles themselves to a smaller size). Milling is generally done by mechanical means, including attrition (which is particle-to-particle collision that results in agglomerate break up or particle shearing), compression (which applies a forces that results in fracturing), and impact (which employs a milling medium or the particles themselves to cause fracturing). Attrition milling equipment includes the wet scrubber (also called the planetary mill or wet attrition mill), which has paddles in water creating vortexes in which the material collides and break up.

Compression mills include the jaw crusher, roller crusher and cone crusher. Impact mills include the ball mill, which has media that tumble and fracture the material. Shaft impactors cause particle-to particle attrition and compression.

- Batching is the process of weighing the oxides according to recipes, and preparing them for mixing and drying.

- Mixing occurs after batching and is performed with various machines, such as dry mixing ribbon mixers (a type of cement mixer), Mueller mixers, and pug mills. Wet mixing generally involves the same equipment.

- Forming is making the mixed material into shapes, ranging from toilet bowls to spark plug insulators. Forming can involve: (1) Extrusion, such as extruding "slugs" to make bricks, (2) Pressing to make shaped parts, (3) Slip casting, as in making toilet bowls, wash basins and ornamentals like ceramic statues. Forming produces a "green" part, ready for drying. Green parts are soft, pliable, and over time will lose shape. Handling the green product will change its shape. For example, a green brick can be "squeezed", and after squeezing it will stay that way.

- Drying is removing the water or binder from the formed material. Spray drying is widely used to prepare powder for pressing operations. Other dryers are tunnel dryers and periodic dryers. Controlled heat is applied in this two-stage process. First, heat removes water. This step needs careful control, as rapid heating causes cracks and surface defects. The dried part is smaller than the green part, and is brittle, necessitating careful handling, since a small impact will cause crumbling and breaking.

- Sintering is where the dried parts pass through a controlled heating process, and the oxides are chemically changed to cause bonding and densification. The fired part will be smaller than the dried part.

Forming Methods

Ceramic forming techniques include throwing, slipcasting, tape casting, freeze-casting, injection moulding, dry pressing, isostatic pressing, hot isostatic pressing (HIP) and others. Methods for forming ceramic powders into complex shapes are desirable in many areas of technology. Such methods are required for producing advanced, high-temperature structural parts such as heat engine components and turbines. Materials other than ceramics which are used in these processes may include: wood, metal, water, plaster and epoxy—most of which will be eliminated upon firing.

These forming techniques are well known for providing tools and other components with dimensional stability, surface quality, high (near theoretical) density and microstructural uniformity. The increasing use and diversity of speciality forms of ceramics adds to the diversity of process technologies to be used.

Thus, reinforcing fibres and filaments are mainly made by polymer, sol-gel, or CVD processes, but melt processing also has applicability. The most widely used speciality form is layered structures, with tape casting for electronic substrates and packages being pre-eminent. Photo-lithography is of increasing interest for precise patterning of conductors and other components for such packaging. Tape casting or forming processes are also of increasing interest for other applications, ranging from open structures such as fuel cells to ceramic composites.

The other major layer structure is coating, where melt spraying is very important, but chemical and physical vapour deposition and chemical (e.g., sol-gel and polymer pyrolysis) methods are all seeing increased use. Besides open structures from formed tape, extruded structures, such as honeycomb catalyst supports, and highly porous structures, including various foams, for example, reticulated foam, are of increasing use.

Densification of consolidated powder bodies continues to be achieved predominantly by (pressureless) sintering. However, the use of pressure sintering by hot pressing is increasing, especially for non-oxides and parts of simple shapes where higher quality (mainly microstructural homogeneity) is needed, and larger size or multiple parts per pressing can be an advantage.

The Sintering Process

The principles of sintering-based methods are simple ("sinter" has roots in the English "cinder"). The firing is done at a temperature below the melting point of the ceramic. Once a roughly-held-together object called a "green body" is made, it is baked in a kiln, where atomic and molecular diffusion processes give rise to significant changes in the primary microstructural features. This includes the gradual elimination of porosity, which is typically accompanied by a net shrinkage and overall densification of the component. Thus, the pores in the object may close up, resulting in a denser product of significantly greater strength and fracture toughness.

Another major change in the body during the firing or sintering process will be the establishment of the polycrystalline nature of the solid. This change will introduce some form of grain size distribution, which will have a significant impact on the ultimate physical properties of the material. The grain sizes will either be associated with the initial particle size, or possibly the sizes of aggregates or particle clusters which arise during the initial stages of processing.

The ultimate microstructure (and thus the physical properties) of the final product will be limited by and subject to the form of the structural template or precursor which is created in the initial stages of chemical synthesis and physical forming. Hence the importance of chemical powder and polymer processing as it pertains to the synthesis of industrial ceramics, glasses and glass-ceramics.

There are numerous possible refinements of the sintering process. Some of the most common involve pressing the green body to give the densification a head start and reduce the sintering time needed. Sometimes organic binders such as polyvinyl alcohol are added to hold the green body together; these burn out during the firing (at 200–350 °C). Sometimes organic lubricants are added during pressing to increase densification. It is common to combine these, and add binders and lubricants to a powder, then press. (The formulation of these organic chemical additives is an art in itself. This is particularly important in the manufacture of high performance ceramics such as those used by the billions for electronics, in capacitors, inductors, sensors, etc.)

A slurry can be used in place of a powder, and then cast into a desired shape, dried and then sintered. Indeed, traditional pottery is done with this type of method, using a plastic mixture worked with the hands. If a mixture of different materials is used together in a ceramic, the sintering temperature is sometimes above the melting point of one minor component – a *liquid phase* sintering. This results in shorter sintering times compared to solid state sintering.

Strength of Ceramics

A material's strength is dependent on its microstructure. The engineering processes to which a material is subjected can alter its microstructure. The variety of strengthening mechanisms that alter the strength of a material include the mechanism of grain boundary strengthening. Thus, although yield strength is maximized with decreasing grain size, ultimately, very small grain sizes make the material brittle. Considered in tandem with the fact that the yield strength is the parameter that predicts plastic deformation in the material, one can make informed decisions on how to increase the strength of a material depending on its microstructural properties and the desired end effect.

The relation between yield stress and grain size is described mathematically by the Hall-Petch equation which is

$$\sigma_y = \sigma_0 + \frac{k_y}{\sqrt{d}}$$

where k_y is the strengthening coefficient (a constant unique to each material), σ_o is a materials constant for the starting stress for dislocation movement (or the resistance of the lattice to dislocation motion), d is the grain diameter, and σ_y is the yield stress.

Theoretically, a material could be made infinitely strong if the grains are made infinitely small. This is, unfortunately, impossible because the lower limit of grain size is a single unit cell of the material. Even then, if the grains of a material are the size of a single unit cell, then the material is in fact amorphous, not crystalline, since there is no long range order, and dislocations can not be defined in an amorphous material. It has been observed experimentally that the microstructure with the highest yield strength is a grain size of about 10 nanometres, because grains smaller than this undergo another yielding mechanism, grain boundary sliding. Producing engineering materials with this ideal grain size is difficult because of the limitations of initial particle sizes inherent to nanomaterials and nanotechnology.

Theory of Chemical Processing

Microstructural Uniformity

In the processing of fine ceramics, the irregular particle sizes and shapes in a typical powder often lead to non-uniform packing morphologies that result in packing density variations in the powder compact. Uncontrolled agglomeration of powders due to attractive van der Waals forces can also give rise to in microstructural inhomogeneities.

Differential stresses that develop as a result of non-uniform drying shrinkage are directly related to the rate at which the solvent can be removed, and thus highly dependent upon the distribution of porosity. Such stresses have been associated with a plastic-to-brittle transition in consolidated bodies, and can yield to crack propagation in the unfired body if not relieved.

In addition, any fluctuations in packing density in the compact as it is prepared for the kiln are often amplified during the sintering process, yielding inhomogeneous densification. Some pores and other structural defects associated with density variations have been shown to play a detrimental role in the sintering process by growing and thus limiting end-point densities. Differential stresses arising from inhomogeneous densification have also been shown to result in the propagation of internal cracks, thus becoming the strength-controlling flaws.

It would therefore appear desirable to process a material in such a way that it is physically uniform with regard to the distribution of components and porosity, rather than using particle size distributions which will maximize the green density. The containment of a uniformly dispersed assembly of strongly interacting particles in suspension requires total control over particle-particle interactions. Monodisperse colloids provide this potential.

Monodisperse powders of colloidal silica, for example, may therefore be stabilized sufficiently to ensure a high degree of order in the colloidal crystal or polycrystalline colloidal solid which results from aggregation. The degree of order appears to be limited by the time and space allowed for longer-range correlations to be established.

Such defective polycrystalline colloidal structures would appear to be the basic elements of submicrometer colloidal materials science, and, therefore, provide the first step in developing a more rigorous understanding of the mechanisms involved in microstructural evolution in inorganic systems such as polycrystalline ceramics.

Self-assembly

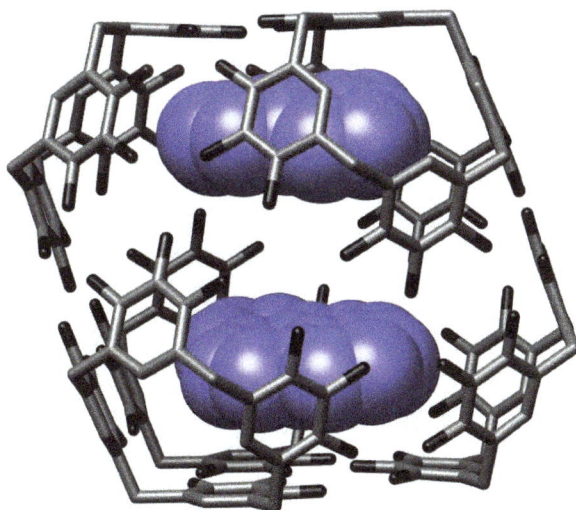

An example of a supramolecular assembly.

Self-assembly is the most common term in use in the modern scientific community to describe the spontaneous aggregation of particles (atoms, molecules, colloids, micelles, etc.) without the influence of any external forces. Large groups of such particles are known to assemble themselves into thermodynamically stable, structurally well-defined arrays, quite reminiscent of one of the 7 crystal systems found in metallurgy and mineralogy (e.g. face-centred cubic, body-centred cubic, etc.). The fundamental difference in equilibrium structure is in the spatial scale of the unit cell (or lattice parameter) in each particular case.

Thus, self-assembly is emerging as a new strategy in chemical synthesis and nanotechnology. Molecular self-assembly has been observed in various biological systems and underlies the formation of a wide variety of complex biological structures. Molecular crystals, liquid crystals, colloids, micelles, emulsions, phase-separated polymers, thin films and self-assembled monolayers all represent examples of the types of highly ordered structures which are obtained using these techniques. The distinguishing feature of these methods is self-organization in the absence of any external forces.

In addition, the principal mechanical characteristics and structures of biological ceramics, polymer composites, elastomers, and cellular materials are being re-evaluated, with an emphasis on bioinspired materials and structures. Traditional approaches focus on design methods of biological materials using conventional synthetic materials. This includes an emerging class of mechanically superior biomaterials based on microstructural features and designs found in nature. The new horizons have been identified in the synthesis of bioinspired materials through processes that are characteristic of biological systems in nature. This includes the nanoscale self-assembly of the components and the development of hierarchical structures.

Ceramic Composites

Substantial interest has arisen in recent years in fabricating ceramic composites. While there is considerable interest in composites with one or more non-ceramic constituents, the greatest attention is on composites in which all constituents are ceramic. These typically comprise two ceramic constituents: a continuous matrix, and a dispersed phase of ceramic particles, whiskers, or short (chopped) or continuous ceramic fibres. The challenge, as in wet chemical processing, is to obtain a uniform or homogeneous distribution of the dispersed particle or fibre phase.

Consider first the processing of particulate composites. The particulate phase of greatest interest is tetragonal zirconia because of the toughening that can be achieved from the phase transformation from the metastable tetragonal to the monoclinic crystalline phase, aka transformation toughening. There is also substantial interest in dispersion of hard, non-oxide phases such as SiC, TiB, TiC, boron, carbon and especially oxide matrices like alumina and mullite. There is also interest too incorporating other ceramic particulates, especially those of highly anisotropic thermal expansion. Examples include Al_2O_3, TiO_2, graphite, and boron nitride.

Silicon carbide single crystal

In processing particulate composites, the issue is not only homogeneity of the size and spatial distribution of the dispersed and matrix phases, but also control of the matrix grain size. However, there is some built-in self-control due to inhibition of matrix grain growth by the dispersed phase. Particulate composites, though generally offer increased resistance to damage, failure, or both, are still quite sensitive to inhomogeneities of composition as well as other processing defects such as pores. Thus they need good processing to be effective.

Particulate composites have been made on a commercial basis by simply mixing powders of the two constituents. Although this approach is inherently limited in the homogeneity that can be achieved, it is the most readily adaptable for existing ceramic production technology. However, other approaches are of interest.

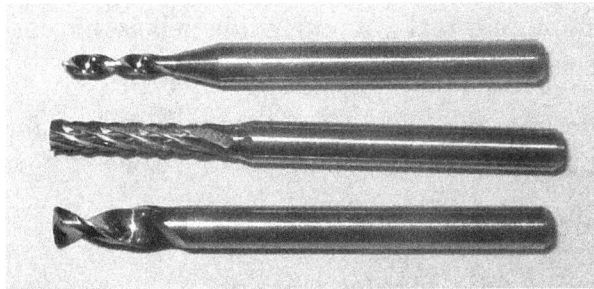

Tungsten carbide milling bits

From the technological standpoint, a particularly desirable approach to fabricating particulate composites is to coat the matrix or its precursor onto fine particles of the dispersed phase with good control of the starting dispersed particle size and the resultant matrix coating thickness. One should in principle be able to achieve the ultimate in homogeneity of distribution and thereby optimize composite performance. This can also have other ramifications, such as allowing more useful composite performance to be achieved in a body having porosity, which might be desired for other factors, such as limiting thermal conductivity.

There are also some opportunities to utilize melt processing for fabrication of ceramic, particulate, whisker and short-fibre, and continuous-fibre composites. Clearly, both particulate and whisker composites are conceivable by solid-state precipitation after solidification of the melt. This can also be obtained in some cases by sintering, as for precipitation-toughened, partially stabilized zirconia. Similarly, it is known that one can directionally solidify ceramic eutectic mixtures and hence obtain uniaxially aligned fibre composites. Such composite processing has typically been limited to very simple shapes and thus suffers from serious economic problems due to high machining costs.

Clearly, there are possibilities of using melt casting for many of these approaches. Potentially even more desirable is using melt-derived particles. In this method, quenching is done in a solid solution or in a fine eutectic structure, in which the particles are then processed by more typical ceramic powder processing methods into a useful body. There have also been preliminary attempts to use melt spraying as a means of forming composites by introducing the dispersed particulate, whisker, or fibre phase in conjunction with the melt spraying process.

Other methods besides melt infiltration to manufacture ceramic composites with long fibre reinforcement are chemical vapour infiltration and the infiltration of fibre preforms with organic precursor, which after pyrolysis yield an amorphous ceramic matrix, initially with a low density. With repeated cycles of infiltration and pyrolysis one of those types of ceramic matrix composites is produced. Chemical vapour infiltration is used to manufacture carbon/carbon and silicon carbide reinforced with carbon or silicon carbide fibres.

Besides many process improvements, the first of two major needs for fibre composites is lower fibre costs. The second major need is fibre compositions or coatings, or composite processing, to reduce degradation that results from high-temperature composite exposure under oxidizing conditions.

Applications

The products of technical ceramics include tiles used in the Space Shuttle program, gas burner nozzles, ballistic protection, nuclear fuel uranium oxide pellets, bio-medical implants, jet engine turbine blades, and missile nose cones.

Its products are often made from materials other than clay, chosen for their particular physical properties. These may be classified as follows:

- Oxides: silica, alumina, zirconia

- Non-oxides: carbides, borides, nitrides, silicides

- Composites: particulate or whisker reinforced matrices, combinations of oxides and non-oxides (e.g. polymers).

Ceramics can be used in many technological industries. One application is the ceramic tiles on NASA's Space Shuttle, used to protect it and the future supersonic space planes from the searing heat of re-entry into the Earth's atmosphere. They are also used widely in electronics and optics. In addition to the applications listed here, ceramics are also used as a coating in various engineering cases. An example would be a ceramic bearing coating over a titanium frame used for an aircraft. Recently the field has come to include the studies of single crystals or glass fibres, in addition to traditional polycrystalline materials, and the applications of these have been overlapping and changing rapidly.

Aerospace

- Engines; Shielding a hot running aircraft engine from damaging other components.
- Airframes; Used as a high-stress, high-temp and lightweight bearing and structural component.
- Missile nose-cones; Shielding the missile internals from heat.
- Space Shuttle tiles
- Space-debris ballistic shields – ceramic fiber woven shields offer better protection to hypervelocity (~7 km/s) particles than aluminium shields of equal weight.
- Rocket nozzles, withstands and focuses the exhaust of the rocket booster.
- Unmanned Air Vehicles; Implications of ceramic engine utilization in aeronautical appli-

cations (such as Unmanned Air Vehicles) may result in enhanced performance characteristics and less operational costs.

Biomedical

A titanium hip prosthesis, with a ceramic head and polyethylene acetabular cup.

- Artificial bone; Dentistry applications, teeth.
- Biodegradable splints; Reinforcing bones recovering from osteoporosis
- Implant material

Electronics

- Capacitors
- Integrated circuit packages
- Transducers
- Insulators

Optical

- Optical fibres, guided lightwave transmission
- Switches
- Laser amplifiers
- Lenses
- Infrared heat-seeking devices

Automotive

- Heat shield
- Exhaust heat management

Biomaterials

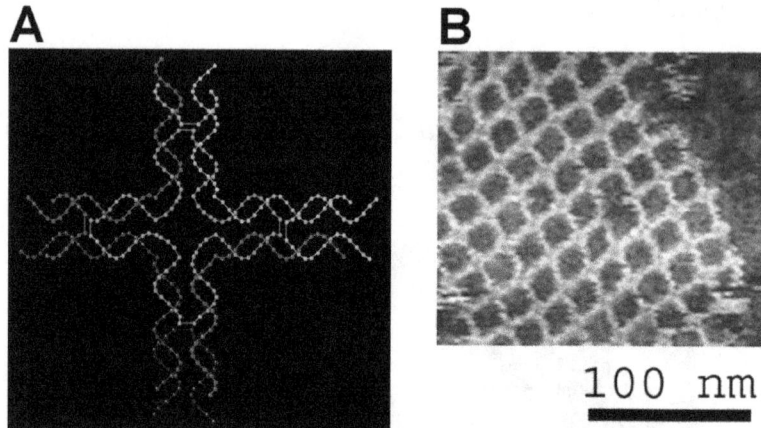

The DNA structure at left (schematic shown) will self-assemble into the structure visualized by atomic force microscopy at right.

Silicification is quite common in the biological world and occurs in bacteria, single-celled organisms, plants, and animals (invertebrates and vertebrates). Crystalline minerals formed in such environment often show exceptional physical properties (e.g. strength, hardness, fracture toughness) and tend to form hierarchical structures that exhibit microstructural order over a range of length or spatial scales. The minerals are crystallized from an environment that is undersaturated with respect to silicon, and under conditions of neutral pH and low temperature (0–40 °C). Formation of the mineral may occur either within or outside of the cell wall of an organism, and specific biochemical reactions for mineral deposition exist that include lipids, proteins and carbohydrates.

Most natural (or biological) materials are complex composites whose mechanical properties are often outstanding, considering the weak constituents from which they are assembled. These complex structures, which have risen from hundreds of million years of evolution, are inspiring the design of novel materials with exceptional physical properties for high performance in adverse conditions. Their defining characteristics such as hierarchy, multifunctionality, and the capacity for self-healing, are currently being investigated.

The basic building blocks begin with the 20 amino acids and proceed to polypeptides, polysaccharides, and polypeptides–saccharides. These, in turn, compose the basic proteins, which are the primary constituents of the 'soft tissues' common to most biominerals. With well over 1000 proteins possible, current research emphasizes the use of collagen, chitin, keratin, and elastin. The 'hard' phases are often strengthened by crystalline minerals, which nucleate and grow in a biomediated environment that determines the size, shape and distribution of individual crystals. The most important mineral phases have been identified as hydroxyapatite, silica, and aragonite. Using the classification of Wegst and Ashby, the principal mechanical characteristics and structures of biological ceramics, polymer composites, elastomers, and cellular materials have been presented. Selected systems in each class are being investigated with emphasis on the relationship between their microstructure over a range of length scales and their mechanical response.

Thus, the crystallization of inorganic materials in nature generally occurs at ambient temperature and pressure. Yet the vital organisms through which these minerals form are capable of consistently producing extremely precise and complex structures. Understanding the processes in which living organisms control the growth of crystalline minerals such as silica could lead to significant advances in the field of materials science, and open the door to novel synthesis techniques for nanoscale composite materials, or nanocomposites.

The iridescent nacre inside a Nautilus shell.

High-resolution SEM observations were performed of the microstructure of the mother-of-pearl (or nacre) portion of the abalone shell. Those shells exhibit the highest mechanical strength and fracture toughness of any non-metallic substance known. The nacre from the shell of the abalone has become one of the more intensively studied biological structures in materials science. Clearly visible in these images are the neatly stacked (or ordered) mineral tiles separated by thin organic sheets along with a macrostructure of larger periodic growth bands which collectively form what scientists are currently referring to as a hierarchical composite structure. (The term hierarchy simply implies that there are a range of structural features which exist over a wide range of length scales).

Future developments reside in the synthesis of bio-inspired materials through processing methods and strategies that are characteristic of biological systems. These involve nanoscale self-assembly of the components and the development of hierarchical structures.

Ceramography

Ceramography is the art and science of preparation, examination and evaluation of ceramic microstructures. Ceramography can be thought of as the metallography of ceramics. The microstructure is the structure level of approximately 0.1 to 100 µm, between the minimum wavelength of visible light and the resolution limit of the naked eye. The microstructure includes most grains, secondary phases, grain boundaries, pores, micro-cracks and hardness microindentions. Most bulk mechanical, optical, thermal, electrical and magnetic properties are significantly affected by the microstructure. The fabrication method and process conditions are generally indicated by the microstructure. The root cause of many ceramic failures is evident in the microstructure. Ceramography is part of the broader field of materialography, which includes all the microscopic techniques of material analysis, such as metallography, petrography and plastography. Ceramography is usually reserved for high-performance ceramics for industrial applications, such as 85–99.9% alumina (Al_2O_3) in Fig. 1, zirconia (ZrO_2), silicon carbide (SiC), silicon nitride (Si_3N_4), and ceramic-matrix composites. It is seldom used on whiteware ceramics such as sanitaryware, wall tiles and dishware.

- Ceramographic microstructures

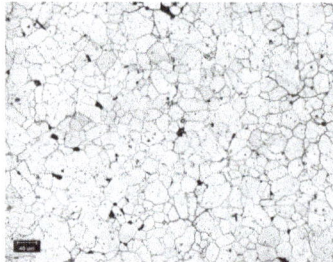

Fig. 1: Thermally etched 99.9% alumina

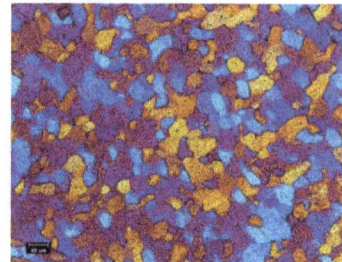

Fig. 2: Thin section of 99.9% alumina

A Brief History of Ceramography

Ceramography evolved along with other branches of materialography and ceramic engineering. Alois de Widmanstätten of Austria etched a meteorite in 1808 to reveal proeutectoid ferrite bands that grew on prior austenite grain boundaries. Geologist Henry Clifton Sorby, the "father of metallography," applied petrographic techniques to the steel industry in the 1860s in Sheffield, England. French geologist Auguste Michel-Lévy devised a chart that correlated the optical properties of minerals to their transmitted color and thickness in the 1880s. Swedish metallurgist J.A. Brinell invented the first quantitative hardness scale in 1900. Smith and Sandland developed the first microindention hardness test at Vickers Ltd. in London in 1922. Swiss-born microscopist A.I. Buehler started the first metallographic equipment manufacturer near Chicago in 1936. Frederick Knoop and colleagues at the National Bureau of Standards developed a less-penetrating (than Vickers) microindention test in 1939. Struers A/S of Copenhagen introduced the electrolytic polisher to metallography in 1943. George Kehl of Columbia University wrote a book that was considered the bible of materialography until the 1980s. Kehl co-founded a group within the Atomic Energy Commission that became the International Metallographic Society in 1967.

Preparation of Ceramographic Specimens

The preparation of ceramic specimens for microstructural analysis consists of five broad steps: sawing, embedding, grinding, polishing and etching. The tools and consumables for ceramographic preparation are available worldwide from metallography equipment vendors and laboratory supply companies.

- Sawing: most ceramics are extremely hard and must be wet-sawed with a circular blade embedded with diamond particles. A metallography or lapidary saw equipped with a low-density diamond blade is usually suitable. The blade must be cooled by a continuous liquid spray.

- Embedding: to facilitate further preparation, the sawed specimen is usually embedded (or mounted or encapsulated) in a plastic disc, 25, 30 or 35 mm in diameter. A thermosetting solid resin, activated by heat and compression, e.g. mineral-filled epoxy, is best for most applications. A castable (liquid) resin such as unfilled epoxy, acrylic or polyester may be used for porous refractory ceramics or microelectronic devices. The castable resins are also available with fluorescent dyes that aid in fluorescence microscopy. The left and right specimens in Fig. 3 were embedded in mineral-filled epoxy. The center refractory in Fig. 3 was embedded in castable, transparent acrylic.

- Grinding is abrasion of the surface of interest by abrasive particles, usually diamond, that are bonded to paper or a metal disc. Grinding erases saw marks, coarsely smooths the surface, and removes stock to a desired depth. A typical grinding sequence for ceramics is one minute on a 240-grit metal-bonded diamond wheel rotating at 240 rpm and lubricated by flowing water, followed by a similar treatment on a 400-grit wheel. The specimen is washed in an ultrasonic bath after each step.

- Polishing is abrasion by free abrasives that are suspended in a lubricant and can roll or slide between the specimen and paper. Polishing erases grinding marks and smooths the specimen to a mirror-like finish. Polishing on a bare metallic platen is called lapping. A typical polishing sequence for ceramics is 5–10 minutes each on 15-, 6- and 1-µm diamond paste or slurry on napless paper rotating at 240 rpm. The specimen is again washed in an ultrasonic bath after each step. The three sets of specimens in Fig. 3 have been sawed, embedded, ground and polished.

- Etching reveals and delineates grain boundaries and other microstructural features that are not apparent on the as-polished surface. The two most common types of etching in ceramography are selective chemical corrosion, and a thermal treatment that causes relief. As an example, alumina can be chemically etched by immersion in boiling concentrated phosphoric acid for 30–60 s, or thermally etched in a furnace for 20–40 min at 1,500 °C (2,730 °F) in air. The plastic encapsulation must be removed before thermal etching. The alumina in Fig. 1 was thermally etched.

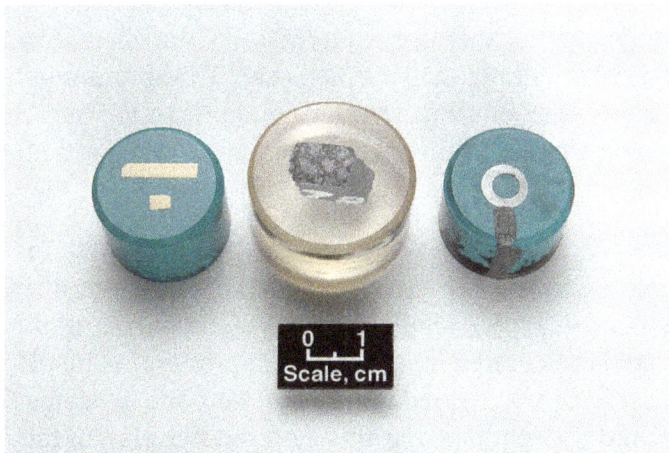

Fig. 3: Embedded, polished ceramographic sections.

Alternatively, non-cubic ceramics can be prepared as thin sections, also known as petrography, for examination by polarized transmitted light microscopy. In this technique, the specimen is sawed to ~1 mm thick, glued to a microscope slide, and ground to a thickness (x) approaching 30 µm. A cover slip is glued onto the exposed surface. The adhesives, such as epoxy or Canada balsam resin, must have approximately the same refractive index ($\eta \approx 1.54$) as glass. Most ceramics have a very small absorption coefficient ($\alpha \approx 0.5$ cm^{-1} for alumina in Fig. 2) in the Beer-Lambert law below, and can be viewed in transmitted light. Cubic ceramics, e.g. yttria-stabilized zirconia and spinel, have the same refractive index in all crystallographic directions and are, therefore, opaque when the microscope's polarizer is 90° out of phase with its analyzer.

$$I_t = I_0 e^{-\alpha x} \quad (\text{(Beer–Lambert eqn)}$$

Ceramographic specimens are electrical insulators in most cases, and must be coated with a conductive ~10-nm layer of metal or carbon for electron microscopy, after polishing and etching. Gold or Au-Pd alloy from a sputter coater or evaporative coater also improves the reflection of visible light from the polished surface under a microscope, by the Fresnel formula below. Bare alumina ($\eta \approx 1.77$, $k \approx 10^{-6}$) has a negligible extinction coefficient and reflects only 8% of the incident light from the microscope, as in Fig. 1. Gold-coated ($\eta \approx 0.82$, $k \approx 1.59$ @ $\lambda = 500$ nm) alumina reflects 44% in air, 39% in immersion oil.

$$R = \frac{I_r}{I_i} = \frac{(\eta_1 - \eta_2)^2 + k^2}{(\eta_1 + \eta_2)^2 + k^2} \quad \text{(Fresnel eqn)}$$

Ceramographic Analysis

Ceramic microstructures are most often analyzed by reflected visible-light microscopy in brightfield. Darkfield is used in limited circumstances, e.g., to reveal cracks. Polarized transmitted light is used with thin sections, where the contrast between grains comes from birefringence. Very fine microstructures may require the higher magnification and resolution of a scanning electron microscope (SEM) or confocal laser scanning microscope (CLSM). The cathodoluminescence microscope (CLM) is useful for distinguishing phases of refractories. The transmission electron microscope (TEM) and scanning acoustic microscope (SAM) have specialty applications in ceramography.

Ceramography is often done qualitatively, for comparison of the microstructure of a component to a standard for quality control or failure analysis purposes. Three common quantitative analyses of microstructures are grain size, second-phase content and porosity. Microstructures are measured by the principles of stereology, in which three-dimensional objects are evaluated in 2-D by projections or cross-sections.

Grain size can be measured by the line-fraction or area-fraction methods of ASTM E112. In the line-fraction methods, a statistical grain size is calculated from the number of grains or grain boundaries intersecting a line of known length or circle of known circumference. In the area-fraction method, the grain size is calculated from the number of grains inside a known area. In each case, the measurement is affected by secondary phases, porosity, preferred orientation, exponential distribution of sizes, and non-equiaxed grains. Image analysis can measure the shape factors of individual grains by ASTM E1382.

Second-phase content and porosity are measured the same way in a microstructure, such as ASTM E562. Procedure E562 is a point-fraction method based on the stereological principle of point fraction = volume fraction, i.e., $P_p = V_v$. Second-phase content in ceramics, such as carbide whiskers in an oxide matrix, is usually expressed as a mass fraction. Volume fractions can be converted to mass fractions if the density of each phase is known. Image analysis can measure porosity, pore-size distribution and volume fractions of secondary phases by ASTM E1245. Porosity measurements do not require etching. Multi-phase microstructures do not require etching if the contrast between phases is adequate, as is usually the case.

Grain size, porosity and second-phase content have all been correlated with ceramic properties such as mechanical strength σ by the Hall–Petch equation. Hardness, toughness, dielectric constant and many other properties are microstructure-dependent.

Microindention Hardness and Toughness

The hardness of a material can be measured in many ways. The Knoop hardness test, a method of microindention hardness, is the most reproducible for dense ceramics. The Vickers hardness test and superficial Rockwell scales (e.g., 45N) can also be used, but tend to cause more surface damage than Knoop. The Brinell test is suitable for ductile metals, but not ceramics. In the Knoop test, a diamond indenter in the shape of an elongated pyramid is forced into a polished (but not etched) surface under a predetermined load, typically 500 or 1000 g. The load is held for some amount of time, say 10 s, and the indenter is retracted. The indention long diagonal (d, μm, in Fig. 4) is measured under a microscope, and the Knoop hardness (HK) is calculated from the load (P, g) and the square of the diagonal length in the equations below. The constants account for the projected area of the indenter and unit conversion factors. Most oxide ceramics have a Knoop hardness in the range of 1000–1500 kg_f/mm^2 (10 − 15 GPa), and many carbides are over 2000 (20 GPa). The method is specified in ASTM C849, C1326 & E384. Microindention hardness is also called microindentation hardness or simply microhardness. The hardness of very small particles and thin films of ceramics, on the order of 100 nm, can be measured by nanoindentation methods that use a Berkovich indenter.

$$HK = 14229 \frac{P}{d^2} \text{ (kg}_f/\text{mm}^2) \text{ and } HK = 139.54 \frac{P}{d^2} \text{ (GPa)}$$

The toughness of ceramics can be determined from a Vickers test under a load of 10 − 20 kg. Toughness is the ability of a material to resist crack propagation. Several calculations have been formulated from the load (P), elastic modulus (E), microindention hardness (H), crack length (c in Fig. 5) and flexural strength (σ). Modulus of rupture (MOR) bars with a rectangular cross-section are indented in three places on a polished surface. The bars are loaded in 4-point bending with the polished, indented surface in tension, until fracture. The fracture normally originates at one of the indentions. The crack lengths are measured under a microscope. The toughness of most ceramics is 2–4 MPa√m, but toughened zirconia is as much as 13, and cemented carbides are often over 20. The toughness-by-indention methods have been discredited recently and are being replaced by more rigorous methods that measure crack growth in a notched beam in flexure.

$$K_{icl} = 0.016 \sqrt{\frac{E}{H}} \frac{P}{(c_0)^{1.5}} \text{ initial crack length}$$

$$K_{isb} = 0.59 \left(\frac{E}{H} \right)^{1/8} [\sigma(P^{1/3})]^{3/4} \text{ indention strength in bending}$$

- Indented Microstructures

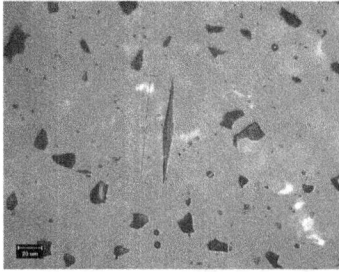

Fig. 4: Knoop indention (P=1kg) in 99.5% alumina

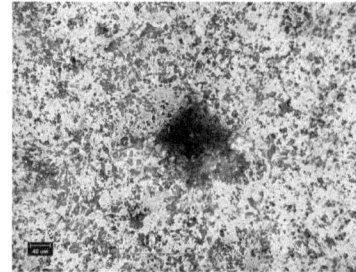

Fig. 5: Toughness indention (P=10kg) in 96% alumina

References

- Black, J. T.; Kohser, R. A. (2012). DeGarmo's materials and processes in manufacturing. Wiley. p. 226. ISBN 978-0-470-92467-9.

- Carter, C. B.; Norton, M. G. (2007). Ceramic materials: Science and engineering. Springer. pp. 20 & 21. ISBN 978-0-387-46271-4.

- Wachtman, John B., Jr. (ed.) (1999) Ceramic Innovations in the 20th century, The American Ceramic Society. ISBN 978-1-57498-093-6.

- von Hippel; A. R. (1954). "Ceramics". Dielectric Materials and Applications. Technology Press (M.I.T.) and John Wiley & Sons. ISBN 1-58053-123-7.

- Harris, D.C., "Materials for Infrared Windows and Domes: Properties and Performance", SPIE PRESS Monograph, Vol. PM70 (Int. Society of Optical Engineers, Bellingham WA, 2009) ISBN 978-0-8194-5978-7

- Brinker, C.J.; Scherer, G.W. (1990). Sol-Gel Science: The Physics and Chemistry of Sol-Gel Processing. Academic Press. ISBN 0-12-134970-5.

Techniques of Ceramic Forming

Freeze gelation helps in the processing of a ceramic object to be fabricated in complex shapes. Alternatively, the techniques used in ceramic forming are freeze-casting and sintering. The aspects elucidated in this chapter are of vital importance and provides a better understanding of ceramic farming.

Ceramic Forming Techniques

Ceramic forming techniques are ways of forming ceramics, which are used to make everyday tableware from teapots, to engineering ceramics such as computer parts. Pottery techniques include the potter's wheel, slipcasting, and many others.

Methods for forming powders of ceramic raw materials into complex shapes are desirable in many areas of technology. For example, such methods are required for producing advanced, high-temperature structural parts such as heat engine components, recuperators and the like from powders of ceramic raw materials. Typical parts produced with this production operation include impellers made from stainless steel, bronze, complex cutting tools, plastic mould tooling, and others. Typical materials used are: wood, metal, water, plaster, epoxy and STLs, silica, and zirconia.

This production operation is well known for providing tools with dimensional stability, surface quality, density and uniformity. For instance, on the slip casting process the cast part is of high concentration of raw materials with little additive, this improves uniformity. But also, the plaster mould draws water from the poured slip to compact and form the casting at the mould surface. This forms a dense cast.

Slip Casting

There are many forming techniques to make ceramics, but one example is slipcasting. This is where slip, liquid clay, is poured into a plaster mould. The water in the slip is drawn out of the slip, leaving an inside layer of solid clay. When this is thick enough, the excess slip can be removed from the mould. When dry, the solid clay can then also be removed. The slip used in slip casting is often liquified with a substance that reduces the need for additional water to soften the slip; this prevents excessive shrinkage which occurs when a piece containing a lot of water dries.

Slip-casting methods provide superior surface quality, density and uniformity in casting high-purity ceramic raw materials over other ceramic casting techniques, such as hydraulic casting, since the cast part is a higher concentration of ceramic raw materials with little additives. A slip is a suspension of fine raw materials powder in a liquid such as water or alcohol with small amounts of secondary materials such as dispersants, surfactants and binders. Pottery slipcasting techniques employ a plaster block or flask mould. The plaster mould draws water from the poured slip to compact and form the casting at the mould surface. This forms a dense cast removing deleterious air gaps and minimizing shrinkage in the final sintering process.

Ceramic Shell Casting

Ceramic shell casting techniques using silica, zirconia and other refractory materials are currently used by the metal parts industry for 'net casting', forming precision shell moulds for molten metal casting. The technique involves a successive wet dipping and dry powder coating or stucco to build up the mould shell layer. The shell casting method in general is known for dimensional stability and is used in many net-casting processes for aerospace and other industries in molten metal casting. Automated facilities use multiple wax patterns on trees, large slurry mixers and fluidic powder beds for automated dipping.

Technical Ceramics

When forming technical ceramic materials from dry powders prepared for processing, the method of forming into the shape required depends upon the method of material preparation and size and shape of the part to be formed. Materials prepared for dry powder forming are most commonly formed by "dry" pressing in mechanical or hydraulic powder compacting presses selected for the necessary force and powder fill depth. Dry powder is automatically discharged into the non-flexible steel or tungsten carbide insert in the die and punches then compact the powder to the shape of the die. If the part is to be large and unable to have pressure transmit suitably for a uniform pressed density then isostatic pressing may be used. When iso-statically pressed the powder takes the shape of a flexible membrane acting as the mould, forming the shape and size of the pressed powder. Isostatic presses can be either high speed, high output type of automatic presses for such parts as ceramic insulators for spark plugs or sand blast nozzles, or slower operating "wet bag" presses that are much more manual in operation but suitable particularly for large machinable blanks or blanks that will be cut or otherwise formed in secondary operations to the final shape.

If technical ceramic parts are needed where the length to diameter ratio is very large, extrusion may be used. There are two types of ceramic extruders one being piston type with hydraulic force pushing a ram that in turn is pushing the ceramic through the loaded material cylinder to and through the die which forms the extrudate. The second type of extruder is a screw, or auger, type where a screw turns forcing the material to and through the die which again shapes the part. In both types of extrusion the raw material must be plasticized to allow and induce the flow of the material in the process.

Complex technical ceramic parts are commonly formed using either the injection moulding process or "hot wax moulding." Both rely on heat sensitive plasticizers to allow material flow into a die. The part is then quickly cooled for removal from the die. Ceramic injection moulding is much like plastic injection moulding using various polymers for plasticizing. Hot wax moulding largely uses paraffin wax.

Other Techniques

There are also several traditional techniques of handbuilding, such as pinching, soft slab, hard slab, and coil construction.

Other techniques involve threading animal or artificial wool fiber through paperclay slip, to build up layers of material. The result can be wrapped over forms or cut, dried and later joined with liquid and soft paperclay.

When forming very thin sheets of ceramic material, "tape casting" is commonly used. This involves pouring the slip (which contains a polymer "binder" to give it strength) onto a moving carrier belt, and then passing it under a stationary "doctor blade" to adjust the thickness. The moving slip is then air dried, and the "tape" thus formed is peeled off the carrier belt, cut into rectangular shapes, and processed further. As many as 100 tape layers, alternating with conductive metal powder layers, can be stacked up. These are then sintered ("fired") to remove the polymer and thus make "multilayer" capacitors, sensors, etc. According to D. W. Richerson of the American Ceramic Society, more than a billion of such capacitors are manufactured every day. (About 100 are in a typical cellular telephone, and about a thousand in a typical automobile.)

Gel casting is another technique used to create engineering ceramics.

Freeze-casting

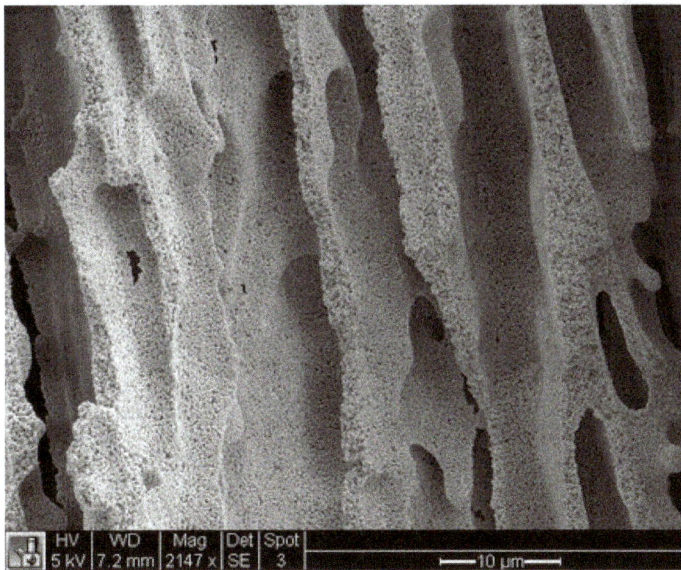

Freeze-Cast alumina that has been partially sintered. The freezing direction in the image is up.

Freeze-casting is a technique that exploits the highly anisotropic solidification behavior of a solvent (generally water) in a well-dispersed slurry to template controllably a directionally porous ceramic. By subjecting an aqueous slurry to a directional temperature gradient, ice crystals will nucleate on one side of the slurry and grow along the temperature gradient. The ice crystals will redistribute the suspended ceramic particles as they grow within the slurry, effectively templating the ceramic.

Once solidification has ended, the frozen, templated ceramic is placed into a freeze-dryer to remove the ice crystals. The resulting green body contains anisotropic macropores in an replica of the sublimated ice crystals and micropores found between the ceramic particles in the walls. This structure is often sintered to consolidate the particulate walls and provide strength to the porous material. The porosity left by the sublimation of solvent crystals is typically between 2 - 200 μm.

Overview

The first observation of cellular structures resulting from the freezing of water goes back over a century, but the first reported instance of freeze-casting, in the modern sense, was in 1954 when Maxwell et al. attempted to fabricate turbosupercharger blades out of refractory powders. They froze extremely thick slips of titanium carbide, producing near-net-shape castings that were easy to sinter and machine. The goal of this work, however, was to make dense ceramics. It was not until 2001, when Fukasawa et al. created directionally porous alumina castings, that the idea of using freeze-casting as a means of creating novel porous structures really took hold. Since that time, research has grown considerably with hundreds of papers of papers coming out within the last decade.

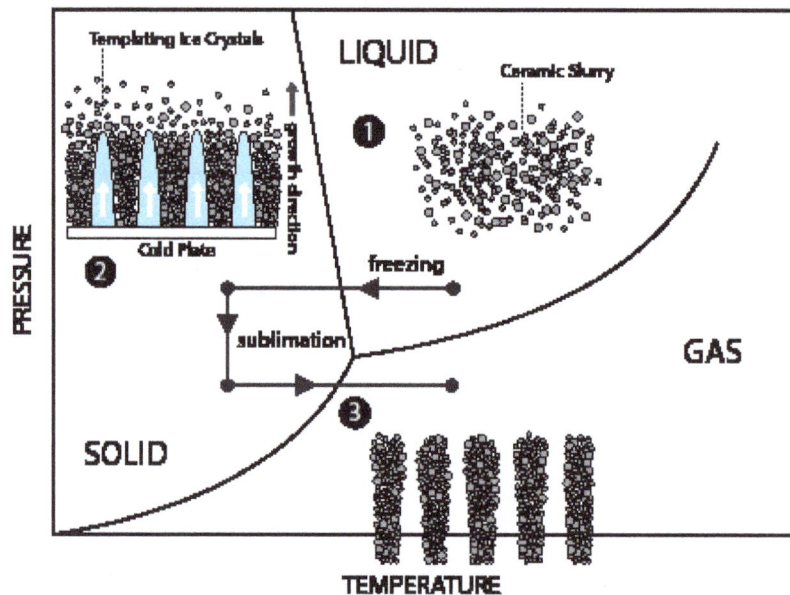

Steps in the freeze-casting process.

Because freeze-casting is a physical process, the techniques developed for one material system can be applied to a wide range of materials. Additionally, due to the inordinate amount of control and broad range of possible porous microstructures that freeze-casting can produce, the technique has found its niche in a number of disparate fields such as tissue scaffolds, photonics, metal-matrix composites, dentistry, materials science, and even food science

There are three possible end results to uni-directionally freezing a suspension of particles. First, the ice-growth proceeds as a planar front, pushing particles in front like a bulldozer pushes a pile of rocks. This scenario usually occurs at very low solidification velocities (< 1 μm s⁻¹) or with extremely fine particles because they can move by Brownian motion away from the front. The resultant structure contains no macroporosity. If one were to increase the solidification speed, the size of the particles or solid loading moderately, the particles begin to interact in a meaningful way with the approaching ice front. The result is typically a lamellar or cellular templated structure whose exact morphology depends on the particular conditions of the system. It is this type of solidification that is targeted for porous materials made by freeze-casting. The third possibility for a freeze-cast structure occurs when particles are given insufficient time to segregate from the suspension, resulting in complete encapsulation of the particles within the ice front. This occurs when the freezing rates are rapid, particle size becomes sufficiently large, or when the solids loading is high enough to hinder particle motion. To ensure templating, the particles must be ejected from the oncoming front. Energetically speaking, this will occur if there is an overall increase in free energy if the particle were to be engulfed $(\Delta\sigma > 0)$.

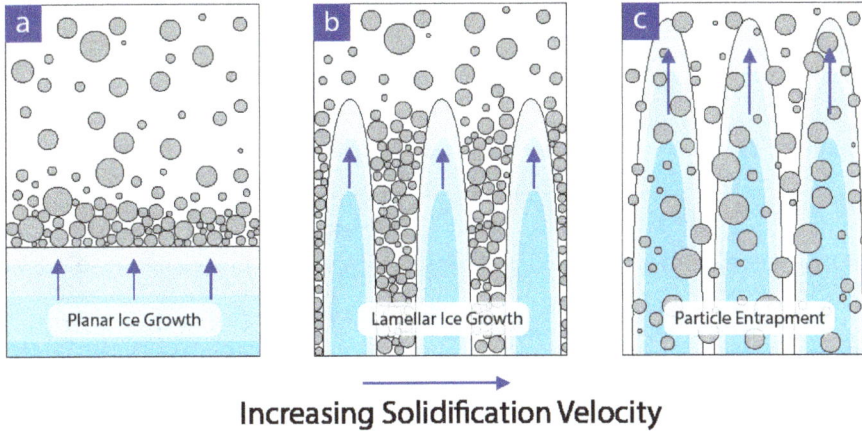

Increasing Solidification Velocity

Depending on the speed of the freezing front, particle size and solids loading there are three pos-sible morphological outcomes: (a) planar front where all particles are pushed ahead of the ice, (b) lamellar/cellular front where ice crystals template particles or (c) particles are engulfed producing no ordering.

$$\Delta\sigma = \sigma_{ps} - (\sigma_{pl} + \sigma_{sl})$$

where $\Delta\sigma$ is the change in free energy of the particle, is the surface potential between the particle and interface, σ_{pl} is the potential between the particle and the liquid phase and σ_{sl} is the surface potential between the solid and liquid phases. This expression is valid at low solidification velocities, when the system is shifted only slightly from equilibrium. At high solidification velocities, kinetics must also be taken into consideration. There will be a liquid film between the front and particle to maintain constant transport of the molecules which are incorporated into the growing crystal. When the front velocity increases, this film thickness (d) will decrease due to increasing drag forces. A critical velocity (v_c) occurs when the film is no longer thick enough to supply the needed molecular supply. At this speed the particle will be engulfed. Most authors express v_c as a function of particle size where $v_c \propto \dfrac{1}{R}$. The transition from a porous R (lamellar) morphology to one where the majority of particles are entrapped occurs at v_c, which was defined by Deville et al. to be:

$$v_c = \frac{\Delta\sigma d}{3\eta R}\left(\frac{a_0}{d}\right)^z$$

where a_0 is the average intermolecular distance of the molecule that is freezing within the liquid, d is the overall thickness of the liquid film, η is the solution viscosity, R is the particle radius and

z is an exponent that can vary from 1 to 5. As expected, we see that v_c decreases as particle radius R goes up.

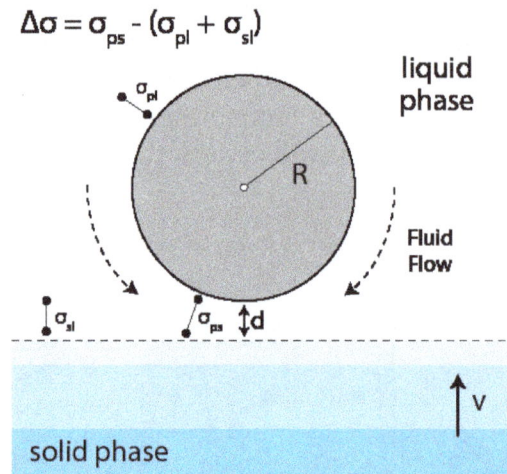

$$\Delta\sigma = \sigma_{ps} - (\sigma_{pl} + \sigma_{sl})$$

Schematic of a particle within the liquid phase interacting with an oncoming solidification front.

Waschkies et al. studied the structure of dilute to concentrated freeze-casts from low (< 1 µm s^{-1}) to extremely high (> 700 µm s^{-1}) solidification velocities. From this study, they were able to generate morphological maps for freeze-cast structures made under various conditions. Maps such as these are excellent for showing general trends, but they are quite specific to the materials system from which they were derived. For most applications where freeze-casts will be used after freezing, binders are needed to supply strength in the green state. The addition of binder can significantly alter the chemistry within the frozen environment, depressing the freezing point and hampering particle motion leading to particle entrapment at speeds far below the predicted v_c. Assuming, however, that we are operating at speeds below VC and above those which produce a planar front, we will achieve some cellular structure with both ice-crystals and walls composed of packed ceramic particles. The morphology of this structure is tied to some variables, but the most influential is the temperature gradient as a function of time and distance along the freezing direction.

Freeze-casts have at least three apparent morphological regions. At the side where freezing initiates is a nearly isotropic region with no visible macropores dubbed the Initial Zone (IZ). Directly after the IZ is the Transition Zone (TZ), where macropores begin to form and align with one another. The pores in this region may appear randomly oriented. The third zone is called the Steady-State Zone (SSZ), macropores in this region are aligned with one another and grow in a regular fashion. Within the SSZ, the structure is defined by a value λ that is the average thickness of a ceramic wall and its adjacent macropore.

Initial Zone: Nucleation and Growth Mechanisms

Although the ability of ice to exclude suspended particles has long been known, the mechanism is still being debated. It was believed initially that during the moments immediately following the nucleation of the ice crystals, particles were ejected from the growing planar ice front, leading to the formation of a constitutionally super-cooled zone directly ahead of the growing ice. This unstable region eventually resulted in perturbations, breaking the planar front into a columnar ice front, a phenomenon better known as a Mullins-Serkerka instability. After the breakdown, the ice crystals grow along the temperature gradient, pushing ceramic particles from the liquid phase aside so that they accumulate between the growing ice crystals. However, recent in-situ X-ray radiography of directionally frozen alumina suspensions reveal a different mechanism.

In-situ testing reveals that freeze-casting is an aggressive growth process. In the moments immediately before nucleation, the suspension is in an unstable super-cooled state. Homogeneous (spatially speaking) nucleation of ice crystals occurs followed by explosive crystal growth in every spatial and crystallographic direction. The initial nucleation and growth steps are so rapid (approaching 800 mm s^{-1}) that all suspended particles are completely engulfed by the oncoming ice front because not enough time is given for particle redistribution, resulting in a structure with anisotropic particle distribution. This step is what provides the initial zone structure.

Transition Zone: A Changing Microstructure

As solidification slows and growth kinetics become rate-limiting, the ice crystals begin to exclude the particles, redistributing them within the suspension. A competitive growth process develops between two crystal populations, those with their basal planes aligned with the thermal gradient (z-crystals) and those that are randomly oriented (r-crystals) giving rise to the start of the TZ.

There are colonies of similarly aligned ice crystals growing throughout the suspension. There are fine lamellae of aligned z-crystals growing with their basal planes aligned with the thermal gradient. The r-crystals appear in this cross-section as platelets but in actuality, they are most similar to columnar dendritic crystals cut along a bias. Within the transition zone, the r-crystals either stop growing or turn into z-crystals that eventually become the predominant orientation, and lead to steady-state growth. There are some reasons why this occurs. For one, during freezing, the growing crystals tend to align with the temperature gradient, as this is the lowest energy configuration and thermodynamically preferential. Aligned growth, however, can mean two different things. Assuming the temperature gradient is vertical, the growing crystal will either be parallel (z-crystal) or perpendicular (r-crystal) to this gradient. A crystal that lays horizontally can still grow in line with the temperature gradient, but it will mean growing on its face rather than its edge. Since the thermal conductivity of ice is so small (1.6 - 2.4 W mK^{-1}) compared with most every other ceramic (ex. Al$_2$O$_3$ = 40 W mK^{-1}), the growing ice will have a significant insulative effect on the localized thermal conditions within the slurry. This can be illustrated using simple resistor elements.

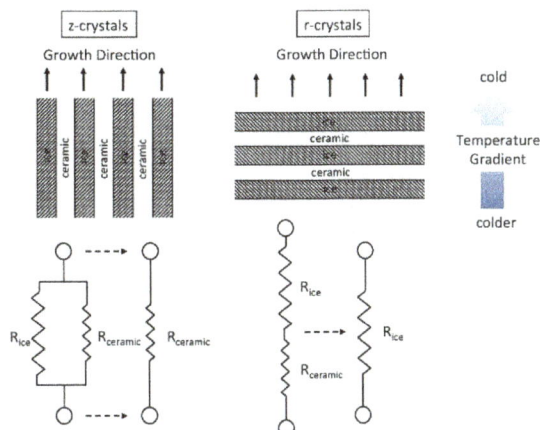

Shows thermal resistance of the two extreme cases of crystallographic alignment.

When ice crystals are aligned with their basal planes parallel to the temperature gradient (z-crystals), they can be represented as two resistors in parallel. The thermal resistance of the ceramic is significantly smaller than that of the ice however, so the apparent resistance can be expressed as the lower $R_{ceramic}$. If the ice crystals are aligned perpendicular to the temperature gradient (r-crystals), they can be approximated as two resistor elements in series. For this case, the R_{ice} is limiting and will dictate the localized thermal conditions. The lower thermal resistance for the z-crystal case leads to lower temperatures and greater heat flux at the growing crystals tips, driving further growth in this direction while, at the same time, the large R_{ice} value hinders the growth of the r-crystals. Each ice crystal growing within the slurry will be some combination of these two scenarios. Thermodynamics dictate that all crystals will tend to align with the preferential temperature gradient causing r-crystals to eventually give way to z-crystals, which can be seen from the following radiographs taken within the TZ.

When z-crystals become the only significant crystal orientation present, the ice-front grows in a steady-state manner except there are no significant changes to the system conditions. It was observed in 2012 that, in the initial moments of freezing, there are dendritic r-crystals that grow 5 - 15 times faster than the solidifying front. These shoot up into the suspension ahead of the main ice front and partially melt back. Interestingly, these crystals stop growing at the point where the TZ will eventually fully transition to the SSZ. Researchers determined that this particular point marks the position where the suspension is in an equilibrium state (i.e. freezing temperature and suspension temperature are equal). We can say then that the size of the initial and transition zones are controlled by the extent of supercooling beyond the already low freezing temperature. If the freeze-casting setup is controlled so that nucleation is favored at only small supercooling, then the TZ will give way to the SSZ sooner.

Steady-state Growth Zone

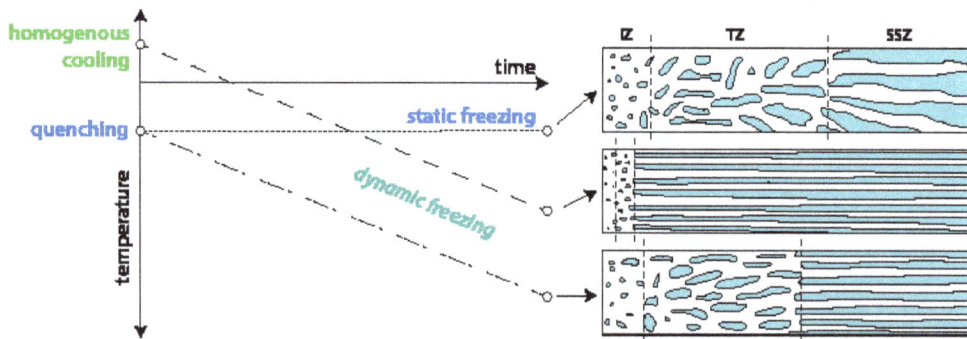

Shows various thermal profiles and their effect on subsequent microstructure of freeze-casts.

The structure in this final region contains long, aligned lamellae that alternate between ice crystals and ceramic walls. The faster a sample is frozen, the finer its solvent crystals (and its eventual macroporosity) will be. Within the SSZ, the normal speeds which are usable for colloidal templating are $10 - 100$ mm s^{-1} leading to solvent crystals typically between 2 mm and 200 mm. Subsequent sublimation of the ice within the SSZ yields a green ceramic preform with porosity in a nearly exact replica of these ice crystals. The microstructure of a freeze-cast within the SSZ is defined by its wavelength (λ) which is the average thickness of a single ceramic wall plus its adjacent macropore. Several publications have reported the effects of solidification kinetics on the microstructures of freeze-cast materials. It has been shown that λ follows an empirical power-law relationship with solidification velocity (v) (Eq. 2.14):

$$\lambda = Av^{-n}$$

Both A and v are used as fitting parameters as currently there is no way of calculating them from first principles, although it is generally believed that A is related to slurry parameters like viscosity and solid loading while n is influenced by particle characteristics.

Controlling the Porous Structure

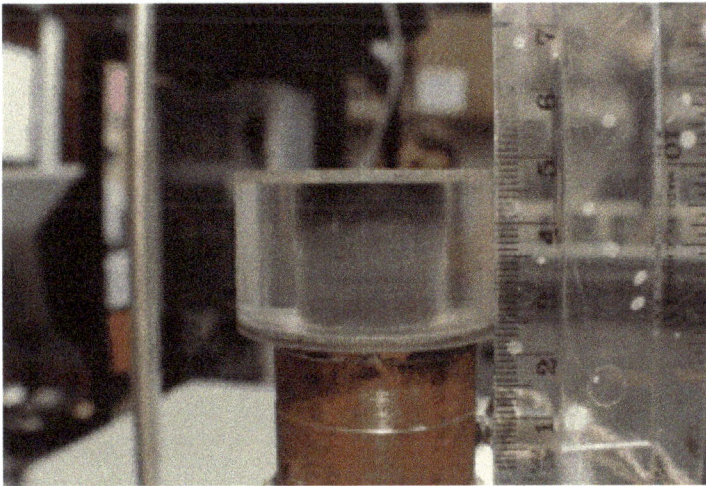

Stop-motion animation of the freeze-casting process.

There are two general categories of tools for architecture a freeze-cast:

1. Chemistry of the System - freezing medium and chosen particulate material(s), any additional binders, dispersants or additives.

2. Operational Conditions - temperature profile, atmosphere, mold material, freezing surface, etc.

Initially, the materials system is chosen based on what sort of final structure is needed. This review has focused on water as the vehicle for freezing, but there are some other solvents that may be used. Notably, camphene, which is an organic solvent that is waxy at room temperature. Freezing of this solution produces highly branched dendritic crystals. Once the materials system is settled on however, the majority of microstructural control comes from external operational conditions such as mold material and temperature gradient.

Controlling Pore Size

The microstructural wavelength (average pore + wall thickness) can be described as a function of the solidification velocity v ($l = Av^{-n}$) where A is dependent on solids loading. There are two ways then that the pore size can be controlled. The first is to change the solidification speed that then alters the microstructural wavelength, or the solids loading can be changed. In doing so, the ratio of pore size to wall size is changed. It is often more prudent to alter the solidification velocity seeing as a minimum solid loading is usually desired. Since microstructural size (λ) is inversely related to the velocity of the freezing front, faster speeds lead to finer structures, while slower speeds produce a coarse microstructure. Controlling the solidification velocity is, therefore, crucial to being able to control the microstructure.

Controlling Pore Shape

Additives can prove highly useful and versatile in changing the morphology of pores. These work by affecting the growth kinetics and microstructure of the ice in addition to the topology of the ice-water interface. Some additives work by altering the phase diagram of the solvent. For example, water and NaCl have a eutectic phase diagram. When NaCl is added into a freeze-casting suspension, the solid ice phase and liquid regions are separated by a zone where both solids and liquids can coexist. This briny region is removed during sublimation, but its existence has a strong effect on the microstructure of the porous ceramic. Other additives work by either altering the interfacial surface energies between the solid/liquid and particle/liquid, changing the viscosity of the suspension, or the degree of undercooling in the system. Studies have been done with glycerol, sucrose, ethanol, coca-cola, acetic acid and more.

Static vs. Dynamic Freezing Profiles

If a freeze casting setup with a constant temperature on either side of the freezing system is used, (static freeze-casting) the front solidification velocity in the SSZ will decrease over time due to the increasing thermal buffer caused by the growing ice front. When this occurs, more time is given for the anisotropic ice crystals to grow perpendicularly to the freezing direction (c-axis) resulting in a structure with ice lamellae that increase in thickness along the length of the sample.

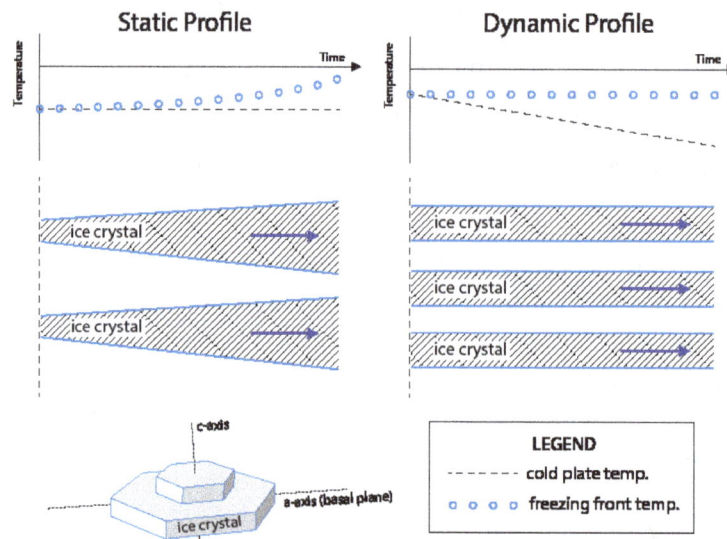

Static and dynamic freezing profiles in the steady-state freezing regime

To ensure highly anisotropic, yet predictable solidification behavior within the SSZ, dynamic freezing patterns are preferred. Using dynamic freezing, the velocity of the solidification front, and, therefore, the ice crystal size, can be controlled with a changing temperature gradient. The increasing thermal gradient counters the effect of the growing thermal buffer imposed by the growing ice front. It has been shown that a linearly decreasing temperature on one side of a freeze-cast will result in near-constant solidification velocity, yielding ice crystals with an almost constant thickness along the SSZ of an entire sample. However, as pointed out by Waschkies et al. even with constant solidification velocity, the thickness of the ice crystals does increase slightly over the course of freezing.

Anisotropy of the Interface Kinetics

Even if the temperature gradient within the slurry is perfectly vertical, it is common to see tilting or curvature of the lamellae as they grow through the suspension. To explain this, it is possible to define two distinct growth directions for each ice crystal. There is the direction determined by the temperature gradient, and the one defined by the preferred growth direction crystallographically speaking. These angles are often at odds with one another, and their balance will describe the tilt of the crystal.

The non-overlapping growth directions also help to explain why dendritic textures are often seen in freeze-casts. This texturing is usually found only on the side of each lamella; the direction of the imposed temperature gradient. The ceramic structure left behind shows the negative image of these dendrites. In 2013, Deville et al. made the interesting observation that the periodicity of these dendrites (tip-to-tip distance) actually seems to be related to the primary crystal thickness.

Particle Packing Effects

Up until now, the focus has been mostly on the structure of the ice itself; the particles are almost an afterthought to the templating process but in fact, the particles can and do play a significant role during freeze-casting. It turns out that particle arrangement also changes as a function of the freezing conditions. For example, researchers have shown that freezing velocity has a marked effect on wall roughness. Faster freezing rates produce rougher walls since particles are given insufficient time to rearrange. This could be of use when developing permeable gas transfer membranes where tortuosity and roughness could impede gas flow. It also turns out that z- and r-crystals do not interact with ceramic particles in the same way. The z-crystals pack particles in the x-y plane while r-crystals pack particles primarily in the z-direction. R-crystals actually pack particles more efficiently than z-crystals and because of this, the area fraction of the particle-rich phase (1 - area fraction of ice crystals) changes as the crystal population shifts from a mixture of z- and r-crystals to only z-crystals. Starting from where ice crystals first begin to exclude particles, marking the beginning of the transition zone, we have a majority of r-crystals and a high value for the particle-rich phase fraction. We can assume that because the solidification speed is still rapid that the particles will not be packed efficiently. As the solidification rate slows down, however, the area fraction of the particle-rich phase drops indicating an increase in packing efficiency. At the same time, the competitive growth process is taking place, replacing r-crystals with z-crystals. At a certain point nearing the end of the transition zone, the particle-rich phase fraction rises sharply since z-crystals are less efficient at packing particles than r-crystals. The apex of this curve marks the point where only z-crystals are present (SSZ). During steady-state growth, after the maximum particle-rich phase fraction is reached, the efficiency of packing increases as steady-state is achieved. In 2011, researchers at Yale University set out to probe the actual spatial packing of particles within the walls. Using small-angle X-ray scattering (SAXS) they characterized the particle size, shape and interparticle spacing of nominally 32 nm silica suspensions that had been freeze-cast at different speeds. Computer simulations indicated that for this system, the particles within the walls should not be touching but rather separated from one another by thin films of ice. Testing however revealed that the particles were, in fact, touching and more than that, they attained a packed morphology that cannot be explained by typical equilibrium densification processes.

Morphological Instabilities

In an ideal world, the spatial concentration of particles within the SSZ would remain constant throughout solidification. As it happens, though, the concentration of particles does change during compression, and this process is highly sensitive to solidification speed. At low freezing rates, Brownian motion takes place, allowing particles to move easily away from the solid-liquid interface and maintain a homogeneous suspension. In this situation, the suspension is always warmer than the solidified portion. At fast solidification speeds, approaching VC, the concentration, and concentration gradient at the solid-liquid interface increases because particles cannot redistribute soon enough. When it has built up enough, the freezing point of the suspension is below the temperature gradient in the solution and morphological instabilities can occur. For situations where the particle concentration bleeds into the diffusion layer, both the actual and freezing temperature dip below the equilibrium freezing temperature creating an unstable system. Often, these situations lead to the formation of what are known as ice lenses.

These morphological instabilities can trap particles, preventing full redistribution and resulting in inhomogeneous distribution of solids along the freezing direction as well as discontinuities in the ceramic walls, creating voids larger than intrinsic pores within the walls of the porous ceramic.

Novel Freeze-Casting Techniques

Freeze-casting can be applied to numerous materials systems including ceramics, polymers and metals. As long as there are particles that may be excluded when the solvent changes phase, a templated structure is possible. Using various novel processing techniques, some authors have demonstrated even greater levels of control made available with freeze-casting. Munch et al. showed that it is possible to control the long-range arrangement and orientation of crystals normal to the growth direction by templating the nucleation surface. This technique works by providing lower energy nucleation sites to control the initial crystal growth and arrangement. The orientation of ice crystals can also be affected by applied electromagnetic fields as was demonstrated in 2010 by Tang et al. Using specialized setups, researchers have been able to create radially aligned freeze-casts tailored for filtration or gas separation applications. Inspired by Nature, scientists have also been able to use coordinating chemicals and cryopreserved to create remarkably distinctive microstructural architectures

Freeze Gelation

Freeze-gelation, is a form of sol-gel processing of ceramics that enables a ceramic object to be fabricated in complex shapes, without the need for high-temperature sintering. The process is similar to freeze-casting.

The process is simple, but the science is, as of 2005, not well understood. The most common process involves the mixing of a silica solution with a filler powder. For example, if we were making a component out of alumina, aluminium oxide, then we would still use a silica sol, but alumina filler powder. The relative amounts used differ, normally between 3 and 4 times more filler than sol is added by weight.

A wetting agent is added, such that the filler powder disperses properly in the sol, which is mostly water. This makes the mixture doughy and stiff. The mixture is, however, highly thixotropic, so that when vibrated it turns liquid. The stiff dough is placed in a mold and the mold vibrated to liquefy the mixture, filling the mold and releasing any trapped air.

The filled mold is then frozen. On freezing, silica precipitates from the sol, forming a gel. This gel holds the filler powder together in something approximating a sintering greenform. The component is then dried in a furnace, leaving the component.

The advantages of freeze-geleation over sintering are essentially cost-based. It doesn't require high pressure equipment or powerful furnaces (drying temperatures are only just above water's boiling point), yet it creates a useful product which takes the shape of the mold very accurately.

History

In terms of being simply a process by which powder can be made into a monolith, freeze casting could be as old as the earth. A material called laminar opaline silica or LOS is believed to be formed by the freeze casting of volcanic ash, some soils containing the required sols to make the gel.

Artificially it is also an old process, having been known and studied for 100 years or more, but never brought to significant industrial application. Lottermoser, a German, wrote a paper on 'das Ausfrieren von Hydrosolen' (the Freezing of Hydrosols) in 1908. Through the 20th century various people have patented techniques using freeze-gelation, most being centred on the use of ceramics as refractory materials. A furnace lining brick, or an investment casting mold, can be easily fabricated using this method.

Recently there was a flurry of interest in freeze-casting at the University of Bath, UK, which led most significantly to two doctoral theses, by J. Laurie in 1995 and by M. Statham in 1998. Taken together in chronological order, these form a good introduction to the technique for the interested party.

Applications

To consider the applications of freeze-casting, we should consider the properties of the freeze-cast component. First, and critically, it is not fully dense. It contains only about 60–70% solid matter, the remainder being air in the form of porosity. This is turn leads to an interesting property of freeze-castings – they are often porous, not merely at the surface, but throughout their thickness. A fluid will penetrate through the pores in the casting and eventually soak through, like a sponge. This is because at porosity percentages above the 'pore percolation threshold', pores link up into continuous channels. The pore percolation threshold depends on the characteristics of the material, but it is normally very roughly around 20%. A 60% dense component has 40% porosity.

As we might expect, this amount of air in the component reduces its strength a lot. Pure, fully dense alumina, for example, is as strong as steel – far stronger if processed carefully – but freeze-cast alumina components are of similar strength to concrete. The freeze-cast component also tends to be brittle, fracturing easily.

It is unlikely then that freeze-cast components could be used structurally (without further processing – more later), but they have other properties that make them useful. They are rather light, with freeze-cast alumina components having a density somewhere in the region of 2.5 g/cm³, similar to aluminium. They are easy and cheap to make, from inexpensive and safe ingredients and using no dangerous equipment. They can take complex shapes, as they are cast, rather than machined. They can also be very large, probably larger than monolithic ceramic components made by any other process. Finally, and crucially, their porosity means that they can be infiltrated by materials with useful properties, or processed with other materials in. For example, the component could be dipped in molten copper, such that the copper is drawn up by capillary action into the porosity, increasing the conductivity of the component vastly. Alternatively, copper powder could be used as a filler powder in place of some alumina to the same end.

Freeze-cast components, in their basic form, are ideal for use as heat-resisting objects. In this way, they can be useful in metalwork, as molds or as substrates for metal spray-forming. However, with suitable post-processing, they could fulfil many other applications, such as silicon chip mounts, or even engine blocks.

Theory

The science is not particularly well understood. It has been known for years that silica sols (also known as colloidal silica, silicic acid, polysilicic acid) will gel when exposed to temperatures around 0 °C (32 °F). The theoretical mechanism is quite simple:

Colloidal silica is produced by the polymerisation of monosilicic acid, $Si(OH)_4$, until the chains of polysilicic acid become so long they form silica particles with hydroxylated surfaces. On freezing of the sol, the silica particles are rejected away from the solidifying interface and forced into the interstices between the ice crystals. Here, they come into contact with each other, and link via the condensation of their surface hydroxyl groups into siloxane bonds. This, happening throughout the sol, forms a gel.

In a filled sol, the ceramic powder is trapped within the gel, and forms a monolith.

Sintering

Sintering is the process of compacting and forming a solid mass of material by heat or pressure without melting it to the point of liquefaction.

Sintering happens naturally in mineral deposits or as a manufacturing process used with metals, ceramics, plastics, and other materials. The atoms in the materials diffuse across the boundaries of the particles, fusing the particles together and creating one solid piece. Because the sintering temperature does not have to reach the melting point of the material, sintering is often chosen as the shaping process for materials with extremely high melting points such as tungsten and molybdenum. The study of sintering in metallurgy powder-related processes is known as powder metallurgy. An example of sintering can be observed when ice cubes in a glass of water adhere to each other, which is driven by the temperature difference between the water and the ice. Examples of pressure-driven sintering are the compacting of snowfall to a glacier, or the forming of a hard snowball by pressing loose snow together.

The word "sinter" comes from the Middle High German *sinter*, a cognate of English "cinder".

10 cm

Clinker nodules produced by sintering

General Sintering

Sintering is effective when the process reduces the porosity and enhances properties such as strength, electrical conductivity, translucency and thermal conductivity; yet, in other cases, it may be useful to increase its strength but keep its gas absorbency constant as in filters or catalysts. During the firing process, atomic diffusion drives powder surface elimination in different stages, starting from the formation of necks between powders to final elimination of small pores at the end of the process.

The driving force for densification is the change in free energy from the decrease in surface area and lowering of the surface free energy by the replacement of solid-vapor interfaces. It forms new but lower-energy solid-solid interfaces with a total decrease in free energy occurring on sintering 1-micrometre particles a 1 cal/g decrease. On a microscopic scale, material transfer is affected by the change in pressure and differences in free energy across the curved surface. If the size of the particle is small (and its curvature is high), these effects become very large in magnitude. The change in energy is much higher when the radius of curvature is less than a few micrometres, which is one of the main reasons why much ceramic technology is based on the use of fine-particle materials.

For properties such as strength and conductivity, the bond area in relation to the particle size is the determining factor. The variables that can be controlled for any given material are the temperature and the initial grain size, because the vapor pressure depends upon temperature. Through time, the particle radius and the vapor pressure are proportional to $(p_0)^{2/3}$ and to $(p_0)^{1/3}$, respectively.

The source of power for solid-state processes is the change in free or chemical potential energy between the neck and the surface of the particle. This energy creates a transfer of material through the fastest means possible; if transfer were to take place from the particle volume or the grain boundary between particles, then there would be particle reduction and pore destruction. The pore elimination occurs faster for a trial with many pores of uniform size and higher porosity where the boundary diffusion distance is smaller. For the latter portions of the process, boundary and lattice diffusion from the boundary become important.

Control of temperature is very important to the sintering process, since grain-boundary diffusion and volume diffusion rely heavily upon temperature, the size and distribution of particles of the material, the materials composition, and often the sintering environment to be controlled.

Ceramic Sintering

Sintering is part of the firing process used in the manufacture of pottery and other ceramic objects. These objects are made from substances such as glass, alumina, zirconia, silica, magnesia, lime, beryllium oxide, and ferric oxide. Some ceramic raw materials have a lower affinity for water and a lower plasticity index than clay, requiring organic additives in the stages before sintering. The general procedure of creating ceramic objects via sintering of powders includes:

- Mixing water, binder, deflocculant, and unfired ceramic powder to form a slurry;

- Spray-drying the slurry;

- Putting the spray dried powder into a mold and pressing it to form a green body (an unsintered ceramic item);

- Heating the green body at low temperature to burn off the binder;

- Sintering at a high temperature to fuse the ceramic particles together.

All the characteristic temperatures associated with phase transformation, glass transitions, and melting points, occurring during a sinterisation cycle of a particular ceramics formulation (i.e., tails and frits) can be easily obtained by observing the expansion-temperature curves during optical dilatometer thermal analysis. In fact, sinterisation is associated with a remarkable shrinkage of the material because glass phases flow once their transition temperature is reached, and start consolidating the powdery structure and considerably reducing the porosity of the material.

Sintering is performed at high temperature. Besides, second and/or third external force (such as pressure, electrical current) could be used. Commonly used second external force is pressure. So, the sintering that performed just using temperature is generally called "pressureless sintering". Pressureless sintering is possible with graded metal-ceramic composites, with a nanoparticle sintering aid and bulk molding technology. A variant used for 3D shapes is called hot isostatic pressing.

To allow efficient stacking of product in the furnace during sintering and prevent parts sticking together, many manufacturers separate ware using ceramic powder separator sheets. These sheets are available in various materials such as alumina, zirconia and magnesia. They are additionally categorized by fine, medium and coarse particle sizes. By matching the material and particle size to the ware being sintered, surface damage and contamination can be reduced while maximizing furnace loading.

Sintering of Metallic Powders

Iron powder

Most, if not all, metals can be sintered. This applies especially to pure metals produced in vacuum which suffer no surface contamination. Sintering under atmospheric pressure requires the use of a protective gas, quite often endothermic gas. Sintering, with subsequent reworking, can produce a great range of material properties. Changes in density, alloying, or heat treatments can alter the physical characteristics of various products. For instance, the Young's Modulus E_n of sintered iron powders remains insensitive to sintering time, alloying, or particle size in the original powder, but depends upon the density of the final product:

$$E_n / E = (D / d)^{3.4}$$

where D is the density, E is Young's modulus and d is the maximum density of iron.

Sintering is static when a metal powder under certain external conditions may exhibit coalescence, and yet reverts to its normal behavior when such conditions are removed. In most cases, the density of a collection of grains increases as material flows into voids, causing a decrease in overall volume. Mass movements that occur during sintering consist of the reduction of total porosity by repacking, followed by material transport due to evaporation and condensation from diffusion. In the final stages, metal atoms move along crystal boundaries to the walls of internal pores, redistributing mass from the internal bulk of the object and smoothing pore walls. Surface tension is the driving force for this movement.

A special form of sintering (which is still considered part of powder metallurgy) is liquid-state sintering in which at least one but not all elements are in a liquid state. Liquid-state sintering is required for making cemented carbide or tungsten carbide.

Sintered bronze in particular is frequently used as a material for bearings, since its porosity allows lubricants to flow through it or remain captured within it. Sintered copper may be used as a wicking structure in certain types of heat pipe construction, where the porosity allows a liquid agent to move through the porous material via capillary action. For materials that have high melting points such as molybdenum, tungsten, rhenium, tantalum, osmium and carbon, sintering is one of the few viable manufacturing processes. In these cases, very low porosity is desirable and can often be achieved.

Sintered metal powder is used to make frangible shotgun shells called breaching rounds, as used by military and SWAT teams to quickly force entry into a locked room. These shotgun shells are designed to destroy door deadbolts, locks and hinges without risking lives by ricocheting or by flying on at lethal speed through the door. They work by destroying the object they hit and then dispersing into a relatively harmless powder.

Sintered bronze and stainless steel are used as filter materials in applications requiring high temperature resistance while retaining the ability to regenerate the filter element. For example, sintered stainless steel elements are employed for filtering steam in food and pharmaceutical applications, and sintered bronze in aircraft hydraulic systems.

Sintering of powders containing precious metals such as silver and gold is used to make small jewelry items.

Advantages

Particular advantages of the powder technology include:

1. Very high levels of purity and uniformity in starting materials

2. Preservation of purity, due to the simpler subsequent fabrication process (fewer steps) that it makes possible

3. Stabilization of the details of repetitive operations, by control of grain size during the input stages

4. Absence of binding contact between segregated powder particles – or "inclusions" (called stringering) – as often occurs in melting processes

5. No deformation needed to produce directional elongation of grains

6. Capability to produce materials of controlled, uniform porosity.

7. Capability to produce nearly net-shaped objects.

8. Capability to produce materials which cannot be produced by any other technology.

9. Capability to fabricate high-strength material like turbine blades.

10. After sintering the mechanical strength to handling becomes higher.

The literature contains many references on sintering dissimilar materials to produce solid/solid-phase compounds or solid/melt mixtures at the processing stage. Almost any substance can be obtained in powder form, through either chemical, mechanical or physical processes, so basically any material can be obtained through sintering. When pure elements are sintered, the leftover powder is still pure, so it can be recycled.

Disadvantages

Particular disadvantages of the powder technology include:

1. 100% sintered (iron ore) can not be charged in the blast furnace.

2. By sintering one cannot create uniform sizes.

3. Micro- and nano-structures produced before sintering are often destroyed.

Plastics Sintering

Plastic materials are formed by sintering for applications that require materials of specific porosity. Sintered plastic porous components are used in filtration and to control fluid and gas flows. Sintered plastics are used in applications requiring wicking properties, such as marking pen nibs. Sintered ultra high molecular weight polyethylene materials are used as ski and snowboard base materials. The porous texture allows wax to be retained within the structure of the base material, thus providing a more durable wax coating.

Liquid Phase Sintering

For materials which are difficult to sinter, a process called liquid phase sintering is commonly used. Materials for which liquid phase sintering is common are Si_3N_4, WC, SiC, and more. Liquid phase sintering is the process of adding an additive to the powder which will melt before the matrix phase. The process of liquid phase sintering has three stages:

- Rearrangement – As the liquid melts capillary action will pull the liquid into pores and also cause grains to rearrange into a more favorable packing arrangement.

- Solution-Precipitation – In areas where capillary pressures are high (particles are close together) atoms will preferentially go into solution and then precipitate in areas of lower chemical potential where particles are not close or in contact. This is called "contact flattening". This densifies the system in a way similar to grain boundary diffusion in solid state sintering. Ostwald ripening will also occur where smaller particles will go into solution preferentially and precipitate on larger particles leading to densification.

- Final Densification – densification of solid skeletal network, liquid movement from efficiently packed regions into pores.

For liquid phase sintering to be practical the major phase should be at least slightly soluble in the liquid phase and the additive should melt before any major sintering of the solid particulate network occurs, otherwise rearrangement of grains will not occur. Liquid phase sintering was successfully applied to improve grain growth of thin semiconductor layers from nanoparticle precursor films.

Electric Current Assisted Sintering

These techniques employ electric currents to drive or enhance sintering. English engineer A. G. Bloxam registered in 1906 the first patent on sintering powders using direct current in vacuum. The primary purpose of his inventions was the industrial scale production of filaments for incandescent lamps by compacting tungsten or molybdenum particles. The applied current was particularly effective in reducing surface oxides that increased the emissivity of the filaments.

In 1913, Weintraub and Rush patented a modified sintering method which combined electric current with pressure. The benefits of this method were proved for the sintering of refractory metals as well as conductive carbide or nitride powders. The starting boron–carbon or silicon–carbon powders were placed in an electrically insulating tube and compressed by two rods which also served as electrodes for the current. The estimated sintering temperature was 2000 °C.

In the United States, sintering was first patented by Duval d'Adrian in 1922. His three-step process aimed at producing heat-resistant blocks from such oxide materials as zirconia, thoria or tantalia. The steps were: (i) molding the powder; (ii) annealing it at about 2500 °C to make it conducting; (iii) applying current-pressure sintering as in the method by Weintraub and Rush.

Sintering that uses an arc produced via a capacitance discharge to eliminate oxides before direct current heating, was patented by G. F. Taylor in 1932. This originated sintering methods employing pulsed or alternating current, eventually superimposed to a direct current. Those techniques have been developed over many decades and summarized in more than 640 patents.

Of these technologies the most well known is resistance sintering (also called hot pressing) and spark plasma sintering, while Electro Sinter Forging is the latest advancement in this field.

Spark Plasma Sintering

In spark plasma sintering (SPS), external pressure and an electric field are applied simultaneously to enhance the densification of the metallic/ceramic powder compacts. This densification uses lower temperatures and shorter amount of time than typical sintering. For a number of years, it was speculated that the existence of sparks or plasma between particles could aid sintering; however, Hulbert and coworkers systematically proved that the electric parameters used during spark plasma sintering make it (highly) unlikely. In light of this, the name "spark plasma sintering" has been rendered obsolete. Terms such as "Field Assisted Sintering Technique" (FAST), "Electric Field Assisted Sintering" (EFAS), and Direct Current Sintering (DCS) have been implemented by the sintering community. Using a DC pulse as the electric current, spark plasma, spark impact pressure, joule heating, and an electrical field diffusion effect would be created.

Electro Sinter Forging

Electro Sinter Forging is an electric current assisted sintering (ecas) technology originated from Capacitor discharge sintering. It is used for the production of diamond metal matrix composites and under evaluation for the production of hard metals, nitinol and other metals and intermetallics. It is characterized by a very low sintering time allowing machines to sinter at the same speed as a compaction press.

Pressureless Sintering

Pressureless sintering is the sintering of a powder compact (sometimes at very high temperatures, depending on the powder) without applied pressure. This avoids density variations in the final component, which occurs with more traditional hot pressing methods.

The powder compact (if a ceramic) can be created by slip casting, injection moulding, and cold isostatic pressing. After pre-sintering, the final green compact can be machined to its final shape before sintered.

Three different heating schedules can be performed with pressureless sintering: constant-rate of heating (CRH), rate-controlled sintering (RCS), and two-step sintering (TSS). The microstructure and grain size of the ceramics may vary depending on the material and method used.

Constant-rate of heating (CRH), also known as temperature-controlled sintering, consists of heating the green compact at a constant rate up to the sintering temperature. Experiments with zirconia have been performed to optimize the sintering temperature and sintering rate for CRH method. Results showed that the grain sizes were identical when the samples were sintered to the same density, proving that grain size is a function of specimen density rather than CRH temperature mode.

In rate-controlled sintering (RCS), the densification rate in the open-porosity phase is lower than in the CRH method. By definition, the relative density, ρ_{rel}, in open-porosity phase is lower than 90%. Although this should prevent separation of pores from grain boundaries, it has been proven statistically that RCS did not produce smaller grain sizes than CRH for alumina, zirconia, and ceria samples.

Two-step sintering (TSS) uses two different sintering temperatures. The first sintering temperature should guarantee a relative density higher than 75% of theoretical sample density. This will remove supercritical pores from the body. The sample will then be cooled down and held at the second sintering temperature until densification is completed. Grains of cubic zirconia and cubic strontium titanate were significantly refined by TSS compared to CRH. However, the grain size changes in other ceramic materials, like tetragonal zirconia and hexagonal alumina, were not statistically significant.

Microwave Sintering

In microwave sintering, heat is generated internally within the material, rather than via radiative heat transfer from an external heat source. Other benefits of microwave sintering are a better heat diffusion, less time needed to reach the sintering temperature, less heating energy required and improvements in the product properties.

As microwaves can only penetrate a short distance in materials with a high conductivity and a high permeability, microwave sintering requires the sample to be delivered in powders with a particle size around the penetration depth of microwaves in the particular material. The sintering process and side-reactions run several times faster during microwave sintering at the same temperature, which results in different properties for the sintered product.

This technique is acknowledged to be quite effective in maintaining fine grains/nano sized grains in sintered bioceramics. Magnesium phosphates and calcium phosphates are the examples which have been processed through microwave sintering technique

Densification, Vitrification and Grain Growth

Sintering in practice is the control of both densification and grain growth. Densification is the act of reducing porosity in a sample thereby making it more dense. Grain growth is the process of grain boundary motion and Ostwald ripening to increase the average grain size. Many properties (mechanical strength, electrical breakdown strength, etc.) benefit from both a high relative density and a small grain size. Therefore, being able to control these properties during processing is of high technical importance. Since densification of powders requires high temperatures, grain growth naturally occurs during sintering. Reduction of this process is key for many engineering ceramics.

For densification to occur at a quick pace it is essential to have (1) an amount of liquid phase that is large in size, (2) a near complete solubility of the solid in the liquid, and (3) wetting of the solid by the liquid. The power behind the densification is derived from the capillary pressure of the liquid phase located between the fine solid particles. When the liquid phase wets the solid particles, each space between the particles becomes a capillary in which a substantial capillary pressure is developed. For submicrometre particle sizes, capillaries with diameters in the range of 0.1 to 1 micrometres develop pressures in the range of 175 pounds per square inch (1,210 kPa) to 1,750 pounds per square inch (12,100 kPa) for silicate liquids and in the range of 975 pounds per square inch (6,720 kPa) to 9,750 pounds per square inch (67,200 kPa) for a metal such as liquid cobalt.

Densification requires constant capillary pressure where just solution-precipitation material transfer would not produce densification. For further densification, additional particle movement while the particle undergoes grain-growth and grain-shape changes occurs. Shrinkage would result when the liquid slips between particles and increase pressure at points of contact causing the material to move away from the contact areas forcing particle centers to draw near each other.

The sintering of liquid-phase materials involves a fine-grained solid phase to create the needed capillary pressures proportional to its diameter and the liquid concentration must also create the required capillary pressure within range, else the process ceases. The vitrification rate is dependent upon the pore size, the viscosity and amount of liquid phase present leading to the viscosity of the overall composition, and the surface tension. Temperature dependence for densification controls the process because at higher temperatures viscosity decreases and increases liquid content. Therefore, when changes to the composition and processing are made, it will affect the vitrification process.

Sintering Mechanisms

Sintering occurs by diffusion of atoms through the microstructure. This diffusion is caused by a gradient of chemical potential – atoms move from an area of higher chemical potential to an area of lower chemical potential. The different paths the atoms take to get from one spot to another are the sintering mechanisms. The six common mechanisms are:

- Surface diffusion – Diffusion of atoms along the surface of a particle
- Vapor transport – Evaporation of atoms which condense on a different surface
- Lattice diffusion from surface – atoms from surface diffuse through lattice
- Lattice diffusion from grain boundary – atom from grain boundary diffuses through lattice
- Grain boundary diffusion – atoms diffuse along grain boundary
- Plastic deformation – dislocation motion causes flow of matter

Also one must distinguish between densifying and non-densifying mechanisms. 1–3 above are non-densifying – they take atoms from the surface and rearrange them onto another surface or part of the same surface. These mechanisms simply rearrange matter inside of porosity and do not cause pores to shrink. Mechanisms 4–6 are densifying mechanisms – atoms are moved from the bulk to the surface of pores thereby eliminating porosity and increasing the density of the sample.

Grain Growth

A grain boundary(GB) is the transition area or interface between adjacent crystallites (or grains) of the same chemical and lattice composition. The adjacent grains do not have the same orientation of the lattice thus giving the atoms in GB shifted positions relative to the lattice in the crystals. Due to the shifted positioning of the atoms in the GB they have a higher energy state when compared with the atoms in the crystal lattice of the grains. It is this imperfection that makes it possible to selectively etch the GBs when one wants the microstructure visible. Striving to minimize its energy leads to the coarsening of the microstructure to reach a metastable state within the specimen. This involves minimizing its GB area and changing its topological structure to minimize its energy. This grain growth can either be normal or abnormal, a normal grain growth is characterized by the uniform growth and size of all the grains in the specimen. Abnormal grain growth is when a few grains grow much larger than the remaining majority.

Grain Boundary Energy/Tension

The atoms in the GB are normally in a higher energy state than their equivalent in the bulk material. This is due to their more stretched bonds, which gives rise to a GB tension σ_{GB}. This extra energy that the atoms possess is called the grain boundary energy, γ_{GB}. The grain will want to minimize this extra energy thus striving to make the grain boundary area smaller and this change requires energy.

"Or, in other words, a force has to be applied, in the plane of the grain boundary and acting along a line in the grain-boundary area, in order to extend the grain-boundary area in the direction of the force. The force per unit length, i.e. tension/stress, along the line mentioned is σGB. On the basis of this reasoning it would follow:

$$\sigma_{GB}dA \text{ (work done)} = \gamma_{GB}dA \text{ (energy change)}$$

with dA as the increase of grain-boundary area per unit length along the line in the grain-boundary area considered." [pg 478]

The GB tension can also be thought of as the attractive forces between the atoms at the surface and the tension between these atoms is due to the fact that there is a larger interatomic distance between them at the surface compared to the bulk (i.e. surface tension). When the surface area becomes bigger the bonds stretch more and the GB tension increases. To counteract this increase in tension there must be a transport of atoms to the surface keeping the GB tension constant. This diffusion of atoms accounts for the constant surface tension in liquids. Then the argument,

$$\sigma_{GB}dA \text{ (work done)} = \gamma_{GB}dA \text{ (energy change)}$$

holds true. For solids, on the other hand, diffusion of atoms to the surface might not be sufficient and the surface tension can vary with an increase in surface area. For a solid, one can derive an expression for the change in Gibbs free energy, dG, upon the change of GB area, dA. dG is given by

$$\sigma_{GB}dA \text{ (work done)} = dG \text{ (energy change)} = \gamma_{GB}dA + Ad\gamma_{GB}$$

which gives

$$\sigma_{GB} = \gamma_{GB} + \frac{Ad\gamma_{GB}}{dA}$$

σ_{GB} is normally expressed in units of $\frac{N}{m}$ while γ_{GB} is normally expressed in units of $\frac{J}{m^2}$ $(J = Nm)$ since they are different physical properties.

Mechanical Equilibrium

In a two-dimensional isotropic material the grain boundary tension would be the same for the grains. This would give angle of 120° at GB junction where three grains meet. This would give the structure a hexagonal pattern which is the metastable state (or mechanical equilibrium) of the 2D specimen. A consequence of this is that to keep trying to be as close to the equilibrium as possible. Grains with fewer sides than six will bend the GB to try keep the 120° angle between each other. This results in a curved boundary with its curvature towards itself. A grain with six sides will, as mentioned, have straight boundaries while a grain with more than six sides will have curved boundaries with its curvature away from itself. A grain with six boundaries (i.e. hexagonal struc- ture) are in a metastable state (i.e. local equilibrium) within the 2D structure. In three dimensions structural details are similar but much more complex and the metastable structure for a grain is a non-regular 14-sided polyhedra with doubly curved faces. In practice all arrays of grains are always unstable and thus always grows until its prevented by a counterforce.

Grains strive to minimize their energy, and a curved boundary has a higher energy than a straight boundary. This means that the grain boundary will migrate towards the curvature. The conse- quence of this is that grains with less than 6 sides will decrease in size while grains with more than 6 sides will increase in size.

Grain growth occurs due to motion of atoms across a grain boundary. Convex surfaces have a higher chemical potential than concave surfaces therefore grain boundaries will move toward their center of curvature. As smaller particles tend to have a higher radius of curvature and this results in smaller grains losing atoms to larger grains and shrinking. This is a process called Ostwald rip- ening. Large grains grow at the expense of small grains. Grain growth in a simple model is found to follow:

$$G^m = G_0^m + Kt$$

Here G is final average grain size, G_o is the initial average grain size, t is time, m is a factor between 2 and 4, and K is a factor given by:

$$K = K_0 e^{\frac{-Q}{RT}}$$

Here Q is the molar activation energy, R is the ideal gas constant, T is absolute temperature, and K_o is a material dependent factor.

Reducing Grain Growth

Solute Ions

If a dopant is added to the material (example: Nd in BaTiO$_3$) the impurity will tend to stick to the grain boundaries. As the grain boundary tries to move (as atoms jump from the convex to concave surface) the change in concentration of the dopant at the grain boundary will impose a drag on the boundary. The original concentration of solute around the grain boundary will be asymmetrical in most cases. As the grain boundary tries to move the concentration on the side opposite of motion will have a higher concentration and therefore have a higher chemical potential. This increased chemical potential will act as a backforce to the original chemical potential gradient that is the reason for grain boundary movement. This decrease in net chemical potential will decrease the grain boundary velocity and therefore grain growth.

Fine second phase particles

If particles of a second phase which are insoluble in the matrix phase are added to the powder in the form of a much finer powder than this will decrease grain boundary movement. When the grain boundary tries to move past the inclusion diffusion of atoms from one grain to the other will be hindered by the insoluble particle. Since it is beneficial for particles to reside in the grain boundaries and they exert a force in opposite direction compared to the grain boundary migration. This effect is called the Zener effect after the man who estimated this drag force to

$$F = \pi r \lambda \sin(2\theta)$$

where r is the radius of the particle and λ the interfacial energy of the boundary if there are N particles per unit volume their volume fraction f is

$$f = \frac{4}{3}\pi r^3 N$$

assuming they are randomly distributed. A boundary of unit area will intersect all particles within a volume of 2r which is 2Nr particles. So the number of particles n intersecting a unit area of grain boundary is:

$$n = \frac{3f}{2\pi r^2}$$

Now assuming that the grains only grow due to the influence of curvature, the driving force of growth is $\frac{2\lambda}{R}$ where (for homogeneous grain structure) R approximates to the mean diameter of the grains. With this the critical diameter that has to be reached before the grains ceases to grow:

$$nF_{max} = \frac{2\lambda}{D_{crit}}$$

This can be reduced to $D_{crit} = \frac{4r}{3f}$ so the critical diameter of the grains is dependent of the size and volume fraction of the particles at the grain boundaries.

It has also been shown that small bubbles or cavities can act as inclusion

More complicated interactions which slow grain boundary motion include interactions of the surface energies of the two grains and the inclusion and are discussed in detail by C.S. Smith.

Natural Sintering in Geology

Petrifying spring in Réotier near Mont-Dauphin, France

In geology a natural sintering occurs when a mineral spring brings about a deposition of chemical sediment or crust, for example as of porous silica.

A sinter is a mineral deposit that presents a porous or vesicular texture; its structure shows small cavities. These may be siliceous deposits or calcareous deposits.

Siliceous sinter is a deposit of opaline or amorphous silica which appears as incrustations near hot springs and geysers. It sometimes forms conical mounds, called geyser cones, but can also form as a terrace. The main agents responsible for the deposition of siliceous sinter are algae and other vegetation in the water. Altering of wall rocks can also form sinters near fumaroles and in the deeper channels of hot springs. Examples of siliceous sinter are geyserite and fiorite. They can be found in many places, including Iceland, El Tatio geothermal field in Chile, New Zealand, and Yellowstone National Park and Steamboat Springs in the USA.

Calcareous sinter is also called tufa, calcareous tufa, or calc-tufa. It is a deposit of calcium carbonate, as with travertine. Called petrifying springs, they are quite common in limestone districts. Their calcareous waters deposit a sintery incrustation on surrounding objects. The precipitation is assisted with mosses and other vegetable structures, thus leaving cavities in the calcareous sinter after they have decayed.

Petrifying spring at Pamukkale, Turkey:

Sintering of Catalysts

Sintering is an important cause for loss of catalyst activity, especially on supported metal catalysts. It decreases the surface area of the catalyst and changes the surface structure. For a porous catalytic surface, the pores may collapse due to sintering, resulting in loss of surface area. Sintering is in general an irreversible process.

Small catalyst particles (which have the highest relative surface areas) and a high reaction temperature are in general both factors that increase the reactivity of a catalyst. However, these factors are also the circumstances under which sintering is occurring. Specific materials may also increase the rate of sintering. On the other hand, by alloying catalysts with other materials, sintering can be reduced. Especially rare earth metals have been shown to reduce sintering of metal catalysts when alloyed.

For many supported metal catalysts, sintering starts to become a significant effect at temperatures over 500 °C (932 °F). Catalysts that operate at higher temperatures, such as a car catalyst, use structural improvements to reduce or prevent sintering. These improvements are in general in the form of a support made from an inert and thermally stable material such as silica, carbon or alumina.

References

- Smallman R. E., Bishop, Ray J (1999). Modern physical metallurgy and materials engineering: science, process, applications. Oxford : Butterworth-Heinemann. ISBN 978-0-7506-4564-5.

- Kang, Suk-Joong L. (2005). Sintering: Densification, Grain Growth, and Microstructure. Elsevier Ltd. pp. 9–18. ISBN 978-0-7506-6385-4.

- Robert W. Cahn, Peter Haasen (1996). Physical Metallurgy (Fourth Edition). pp. 2399–2500. ISBN 978-0-444-89875-3.

- C. Barry Carter, M. Grant Norton (2007). Ceramic Materials: Science and Engineering. Springer Science+Business Media, LLC. pp. 427–443. ISBN 978-0-387-46270-7.

- G. Kuczynski (6 December 2012). Sintering and Catalysis. Springer Science & Business Media. ISBN 978-1-4684-0934-5.

- I. Chorkendorff; J. W. Niemantsverdriet (6 March 2006). Concepts of Modern Catalysis and Kinetics. John Wiley & Sons. ISBN 978-3-527-60564-4.

- Tuan, W.H.; Guo, J.K. (2004). "Multiphased ceramic materials: processing and potential". Springer. ISBN 3-540-40516-X.

- Kingery, W. David; Bowen, H. K.; Uhlmann, Donald R. (April 1976). "Introduction to Ceramics" (2nd ed.). John Wiley & Sons, Academic Press. ISBN 0-471-47860-1.

- Greene, Eric S. (20 October 2006). "Mass transfer in graded microstructure solid oxide fuel cell electrodes". Journal of Power Sources - Volume 161, Issue 1. Elsevier B.V. pp. 225–231. Retrieved 19 May 2016.

- Deville, Sylvain (April 2007). "Ice-templated porous alumina structures". Acta Materialia - Volume 55, Issue 6. Elsevier B.V. pp. 1965–1974. Retrieved 19 May 2016.

- Deville, Sylvain (March 2008). "Freeze-Casting of Porous Ceramics: A Review of Current Achievements and Issues". Advanced Engineering Materials - Vol 10 Issue 3. John Wiley & Sons, Inc. pp. 155–169. Retrieved 19 May 2016.

- Lottermoser, A. (October–December 1908). "Über das Ausfrieren von Hydrosolen". Berichte der deutschen chemischen Gesellschaft - Volume 41, Issue 3. John Wiley & Sons, Inc. pp. 3976–3979. Retrieved 19 May 2016.

Innovations in Ceramics

This section focuses on the innovations in ceramics. Some of these innovations are zirconium dioxide, boron carbide, silicon nitride, nanoceramic and coade stones. The basic purpose of zirconium dioxide is the production of ceramics whereas silicon nitride bearings are full ceramic bearings. Silicon nitride ceramics has good shock resistance as compared to other ceramics. This section has been carefully written to provide an in-depth understanding of the use of ceramics.

Zirconium Dioxide

Zirconium dioxide (ZrO_2), sometimes known as zirconia, is a white crystalline oxide of zirconium. Its most naturally occurring form, with a monoclinic crystalline structure, is the mineral baddeleyite. A dopant stabilized cubic structured zirconia, cubic zirconia, is synthesized in various colours for use as a gemstone and a diamond simulant.

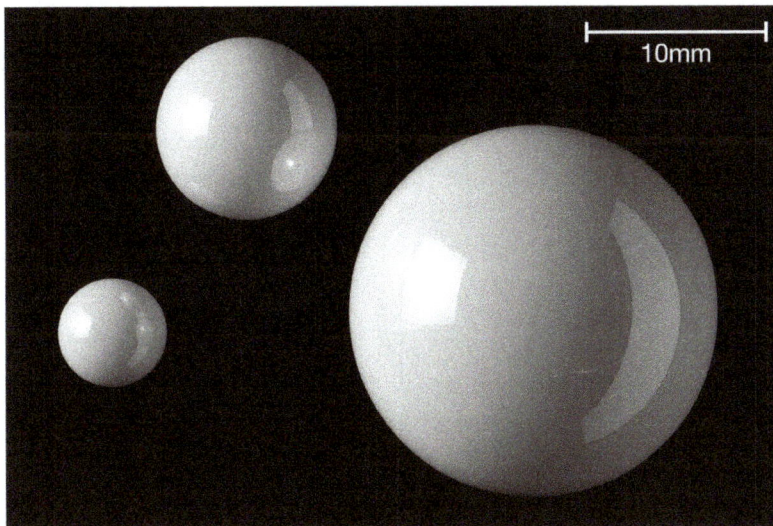

Bearing balls

Production, Chemical Properties, Occurrence

Zirconia is produced by calcining zirconium compounds, exploiting its high thermal stability.

Structure

Three phases are known: monoclinic <1,170 °C, tetragonal 1,170–2,370 °C, and cubic >2,370 °C. The trend is for higher symmetry at higher temperatures, as is usually the case. A few percentage of the oxides of calcium or yttrium stabilize the cubic phase. The very rare mineral tazheranite

(Zr,Ti,Ca)O$_2$ is cubic. Unlike TiO$_2$, which features six-coordinate Ti in all phases, monoclinic zirconia consists of seven-coordinate zirconium centres. This difference is attributed to the larger size of Zr atom relative to the Ti atom.

Chemical Reactions

Zirconia is chemically unreactive. It is slowly attacked by concentrated hydrofluoric acid and sulfuric acid. When heated with carbon, it converts to zirconium carbide. When heated with carbon in the presence of chlorine, it converts to zirconium tetrachloride. This conversion is the basis for the purification of zirconium metal and is analogous to the Kroll process.

Engineering Properties

Zirconium dioxide is one of the most studied ceramic materials. ZrO$_2$ adopts a monoclinic crystal structure at room temperature and transitions to tetragonal and cubic at higher temperatures. The volume expansion caused by the cubic to tetragonal to monoclinic transformation induces large stresses, and these stresses cause ZrO$_2$ to crack upon cooling from high temperatures. When the zirconia is blended with some other oxides, the tetragonal and/or cubic phases are stabilized. Effective dopants include magnesium oxide (MgO), yttrium oxide (Y$_2$O$_3$, yttria), calcium oxide (CaO), and cerium(III) oxide (Ce$_2$O$_3$).

Zirconia is often more useful in its phase 'stabilized' state. Upon heating, zirconia undergoes disruptive phase changes. By adding small percentages of yttria, these phase changes are eliminated, and the resulting material has superior thermal, mechanical, and electrical properties. In some cases, the tetragonal phase can be metastable. If sufficient quantities of the metastable tetragonal phase is present, then an applied stress, magnified by the stress concentration at a crack tip, can cause the tetragonal phase to convert to monoclinic, with the associated volume expansion. This phase transformation can then put the crack into compression, retarding its growth, and enhancing the fracture toughness. This mechanism is known as transformation toughening, and significantly extends the reliability and lifetime of products made with stabilized zirconia.

The ZrO$_2$ band gap is dependent on the phase (cubic, tetragonal, monoclinic, or amorphous) and preparation methods, with typical estimates from 5–7 eV (0.80–1.12 aJ).

A special case of zirconia is that of tetragonal zirconia polycrystal, or TZP, which is indicative of polycrystalline zirconia composed of only the metastable tetragonal phase.

Uses

The main use of zirconia is in the production of ceramics, with other uses including as a protective coating on particles of titanium dioxide pigments, as a refractory material, in insulation, abrasives and enamels. Stabilized zirconia is used in oxygen sensors and fuel cell membranes because it has the ability to allow oxygen ions to move freely through the crystal structure at high temperatures. This high ionic conductivity (and a low electronic conductivity) makes it one of the most useful electroceramics. Zirconium dioxide is also used as the solid electrolyte in electrochromic devices.

Zirconia is a precursor to the electroceramic lead zirconate titanate (*PZT*), which is a high-K dielectric, which is found in myriad components.

Niche Uses

The very low thermal conductivity of cubic phase of zirconia also has led to its use as a thermal barrier coating, or TBC, in jet and diesel engines to allow operation at higher temperatures. Thermodynamically, the higher the operation temperature of an engine, the greater the possible efficiency. Another low thermal conductivity use is a ceramic fiber insulation for crystal growth furnaces, fuel cell stack insulation and infrared heating systems.

This material is also used in dentistry in the manufacture of 1) subframes for the construction of dental restorations such as crowns and bridges, which are then veneered with a conventional feldspathic porcelain for aesthetic reasons, or of 2) strong, extremely durable dental prostheses constructed entirely from monolithic zirconia, with limited but constantly improving aesthetics.

Zirconia is used to make ceramic knives. Because of its hardness, zirconia based cutlery stays sharp longer than a stainless steel equivalent.

Due to its infusibility and brilliant luminosity when incandescent, it was used as an ingredient of sticks for limelight.

Zirconia has been proposed to electrolyze carbon monoxide and oxygen from the atmosphere of Mars to provide both fuel and oxidizer that could be used as a store of chemical energy for use with surface transportation on Mars. Carbon monoxide/oxygen engines have been suggested for early surface transportation use as both carbon monoxide and oxygen can be straightforwardly produced by zirconia electrolysis without requiring use of any of the Martian water resources to obtain hydrogen, which would be needed for the production of methane or any hydrogen-based fuels.

Zirconia is also a potential high-k dielectric material with potential applications as an insulator in transistors.

Zirconia is also employed in the deposition of optical coatings; it is a high-index material usable from the near-UV to the mid-IR, due to its low absorption in this spectral region. In such applications, it is typically deposited by PVD.

An example of zirconium dioxide (ZrO_2) use in a consumer product is a line of the Omega Speedmaster Moonwatch collection, where the watch housings feature laser-engraved "ZrO_2" insignia.

Diamond Simulant

Single crystals of the cubic phase of zirconia are commonly used as diamond simulant in jewellery. Like diamond, cubic zirconia has a cubic crystal structure and a high index of refraction. Visually discerning a good quality cubic zirconia gem from a diamond is difficult, and most jewellers will have a thermal conductivity tester to identify cubic zirconia by its low thermal conductivity (diamond is a very good thermal conductor). This state of zirconia is commonly called *cubic zirconia*,

CZ, or *zircon* by jewellers, but the last name is not chemically accurate. Zircon is actually the mineral name for naturally occurring zirconium silicate ($ZrSiO_4$).

Brilliant-cut cubic zirconia

Boron Carbide

Boron carbide (chemical formula approximately B_4C) is an extremely hard boron–carbon ceramic, and covalent material used in tank armor, bulletproof vests, engine sabotage powders, as well as numerous industrial applications. With a Vickers Hardness of >30 MPa, it is one of the hardest known materials, behind cubic boron nitride and diamond.

Boron carbide was discovered in 19th century as a by-product of reactions involving metal borides, however, its chemical formula was unknown. It was not until the 1930s that the chemical composition was estimated as B_4C. There remained, however, controversy as to whether or not the material had this exact 4:1 stoichiometry, as in practice the material is always slightly carbon-deficient with regard to this formula, and X-ray crystallography shows that its structure is highly complex, with a mixture of C-B-C chains and B_{12} icosahedra. These features argued against a very simple exact B_4C empirical formula. Because of the B_{12} structural unit, the chemical formula of "ideal" boron carbide is often written not as B_4C, but as $B_{12}C_3$, and the carbon deficiency of boron carbide described in terms of a combination of the $B_{12}C_3$ and $B_{12}CBC$ units.

The ability of boron carbide to absorb neutrons without forming long-lived radionuclides makes it attractive as an absorbent for neutron radiation arising in nuclear power plants and from anti-personnel neutron bombs. Nuclear applications of boron carbide include shielding, control rod and shut down pellets. Within control rods, boron carbide is often powdered, to increase its surface area.

Crystal Structure

Boron carbide has a complex crystal structure typical of icosahedron-based borides. There, B_{12} icosahedra form a rhombohedral lattice unit (space group: *R3m* (No. 166), lattice constants: $a = 0.56$ nm and $c = 1.212$ nm) surrounding a C-B-C chain that resides at the center of the unit cell, and both carbon atoms bridge the neighboring three icosahedra. This structure is layered: the B_{12} icosahedra and bridging carbons form a network plane that spreads parallel to the *c*-plane and stacks along the *c*-axis. The lattice has two basic structure units – the B_{12} icosahedron and the B_6 octahedron. Because of the small size of the B_6 octahedra, they cannot interconnect. Instead, they bond to the B_{12} icosahedra in the neighboring layer, and this decreases bonding strength in the c-plane.

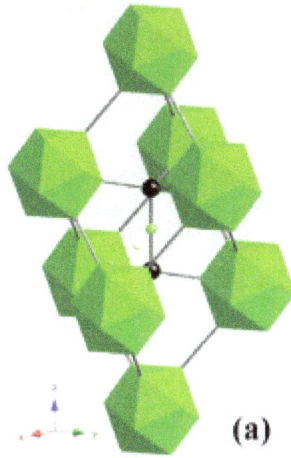

Unit cell of B_4C. The green sphere and icosahedra consist of boron atoms, and black spheres are carbon atoms.

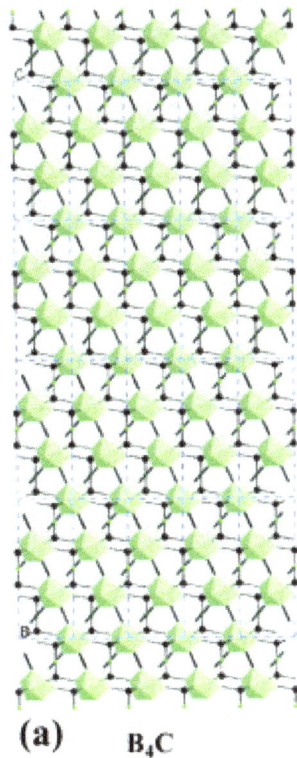

Fragment of the B_4C crystal structure.

Because of the B_{12} structural unit, the chemical formula of "ideal" boron carbide is often written not as B_4C, but as $B_{12}C_3$, and the carbon deficiency of boron carbide described in terms of a combination of the $B_{12}C_3$ and $B_{12}C_2$ units. Some studies indicate the possibility of incorporation of one or more carbon atoms into the boron icosahedra, giving rise to formulas such as $(B_{11}C)CBC = B_4C$ at the carbon-heavy end of the stoichiometry, but formulas such as $B_{12}(CBB) = B_{14}C$ at the boron-rich end. "Boron carbide" is thus not a single compound, but a family of compounds of different compositions. A common intermediate, which approximates a commonly found ratio of elements, is

B_{12}(CBC) = $B_{6.5}$C. Quantum mechanical calculations have demonstrated that configurational disorder between boron and carbon atoms on the different positions in the crystal determines several of the materials properties. In particular the crystal symmetry of the B_4C composition and the non-metallic electrical character of the $B_{13}C_2$ composition.

Properties

Boron carbide is known as a robust material having high hardness, high cross section for absorption of neutrons (i.e. good shielding properties against neutrons), stability to ionizing radiation and most chemicals. Its Vickers hardness (38 GPa), Elastic Modulus (460 GPa) and fracture toughness (3.5 MPa·m$^{1/2}$) approach the corresponding values for diamond (1150 GPa and 5.3 MPa·m$^{1/2}$).

As of 2015, boron carbide is the third hardest substance known, after diamond and cubic boron nitride, earning it the nickname "black diamond".

Semiconductor Properties

Boron carbide is a semiconductor, with electronic properties dominated by hopping-type transport. The energy band gap depends on composition as well as the degree of order. The band gap is estimated at 2.09 eV, with multiple mid-bandgap states which complicate the photoluminescence spectrum. The material is typically p-type.

Preparation

Boron carbide was first synthesized by Henri Moissan in 1899, by reduction of boron trioxide either with carbon or magnesium in presence of carbon in an electric arc furnace. In the case of carbon, the reaction occurs at temperatures above the melting point of B_4C and is accompanied by liberation of large amount of carbon monoxide:

$$2\,B_2O_3 + 7\,C \rightarrow B_4C + 6\,CO$$

If magnesium is used, the reaction can be carried out in a graphite furnace, and the magnesium byproducts are removed by treatment with acid.

Uses

- Padlocks
- Personal and vehicle anti-ballistic armor plating.
- Grit blasting nozzles.
- High-pressure water jet cutter nozzles.
- Scratch and wear resistant coatings.
- Cutting tools and dies.
- Abrasives.

- Neutron absorber in nuclear reactors.

- Metal matrix composites.

- High energy fuel for solid fuel Ramjets.

- In brake linings of vehicles

Silicon Nitride

Silicon nitride is a chemical compound of the elements silicon and nitrogen, with the formula Si_3N_4. It is a white, high-melting-point solid that is relatively chemically inert, being attacked by dilute HF and hot H_2SO_4. It is very hard (8.5 on the mohs scale). It is the most thermodynamically stable of the silicon nitrides. Hence, Si_3N_4 is the most commercially important of the silicon nitrides and is generally understood as what is being referred to where the term "silicon nitride" is used.

Production

The material is prepared by heating powdered silicon between 1300 °C and 1400 °C in an atmosphere of nitrogen:

$$3\,Si + 2\,N_2 \rightarrow Si_3N_4$$

The silicon sample weight increases progressively due to the chemical combination of silicon and nitrogen. Without an iron catalyst, the reaction is complete after several hours (~7), when no further weight increase due to nitrogen absorption (per gram of silicon) is detected. In addition to Si_3N_4, several other silicon nitride phases (with chemical formulas corresponding to varying degrees of nitridation/Si oxidation state) have been reported in the literature, for example, the gaseous disilicon mononitride (Si_2N); silicon mononitride (SiN), and silicon sesquinitride (Si_2N_3), each of which are stoichiometric phases. As with other refractories, the products obtained in these high-temperature syntheses depends on the reaction conditions (e.g. time, temperature, and starting materials including the reactants and container materials), as well as the mode of purification. However, the existence of the sesquinitride has since come into question.

It can also be prepared by diimide route:

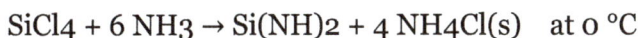

$$SiCl_4 + 6\,NH_3 \rightarrow Si(NH)_2 + 4\,NH_4Cl(s) \quad \text{at 0 °C}$$

$$3Si(NH)_2 \rightarrow Si_3N_4 + N_2 + 3\,H_2(g) \quad \text{at 1000 °C}$$

Carbothermal reduction of silicon dioxide in nitrogen atmosphere at 1400–1450 °C has also been examined:

$$3\,SiO_2 + 6\,C + 2\,N_2 \rightarrow Si_3N_4 + 6\,CO$$

The nitridation of silicon powder was developed in the 1950s, following the "rediscovery" of silicon nitride and was the first large-scale method for powder production. However, use of low-purity raw silicon caused contamination of silicon nitride by silicates and iron. The diimide decomposition results in amorphous silicon nitride, which needs further annealing under nitrogen at 1400–1500 °C to convert it to crystalline powder; this is now the second-most important route for commercial production. The carbothermal reduction was the earliest used method for silicon nitride production and is now considered as the most-cost-effective industrial route to high-purity silicon nitride powder.

Electronic-grade silicon nitride films are formed using chemical vapor deposition (CVD), or one of its variants, such as plasma-enhanced chemical vapor deposition (PECVD):

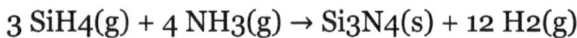

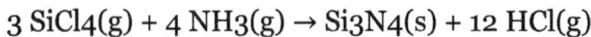

$$3\ SiH_4(g) + 4\ NH_3(g) \rightarrow Si_3N_4(s) + 12\ H_2(g)$$

$$3\ SiCl_4(g) + 4\ NH_3(g) \rightarrow Si_3N_4(s) + 12\ HCl(g)$$

$$3\ SiCl_2H_2(g) + 4\ NH_3(g) \rightarrow Si_3N_4(s) + 6\ HCl(g) + 6\ H_2(g)$$

For deposition of silicon nitride layers on semiconductor (usually silicon) substrates, two methods are used:

1. Low pressure chemical vapor deposition (LPCVD) technology, which works at rather high temperature and is done either in a vertical or in a horizontal tube furnace, or

2. Plasma-enhanced chemical vapor deposition (PECVD) technology, which works at rather low temperature and vacuum conditions.

The lattice constants of silicon nitride and silicon are different. Therefore, tension or stress can occur, depending on the deposition process. Especially when using PECVD technology this tension can be reduced by adjusting deposition parameters.

Silicon nitride nanowires can also be produced by sol-gel method using carbothermal reduction followed by nitridation of silica gel, which contains ultrafine carbon particles. The particles can be produced by decomposition of dextrose in the temperature range 1200–1350 °C. The possible synthesis reactions are:

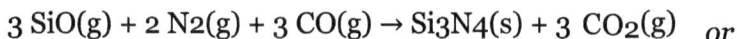

$$SiO_2(s) + C(s) \rightarrow SiO(g) + CO(g) \quad and$$

$$3\ SiO(g) + 2\ N_2(g) + 3\ CO(g) \rightarrow Si_3N_4(s) + 3\ CO_2(g) \quad or$$

$$3\ SiO(g) + 2\ N_2(g) + 3\ C(s) \rightarrow Si_3N_4(s) + 3\ CO(g).$$

Processing

Silicon nitride is difficult to produce as a bulk material—it cannot be heated over 1850 °C, which is well below its melting point, due to dissociation to silicon and nitrogen. Therefore, application of conventional hot press sintering techniques is problematic. Bonding of silicon nitride powders can be achieved at lower temperatures through adding additional materials (sintering aids or "binders") which commonly induce a degree of liquid phase sintering. A cleaner alternative is to use spark plasma sintering where heating is conducted very rapidly (seconds) by passing pulses of electric current through the compacted powder. Dense silicon nitride compacts have been obtained by this techniques at temperatures 1500–1700 °C.

Crystal Structure and Properties

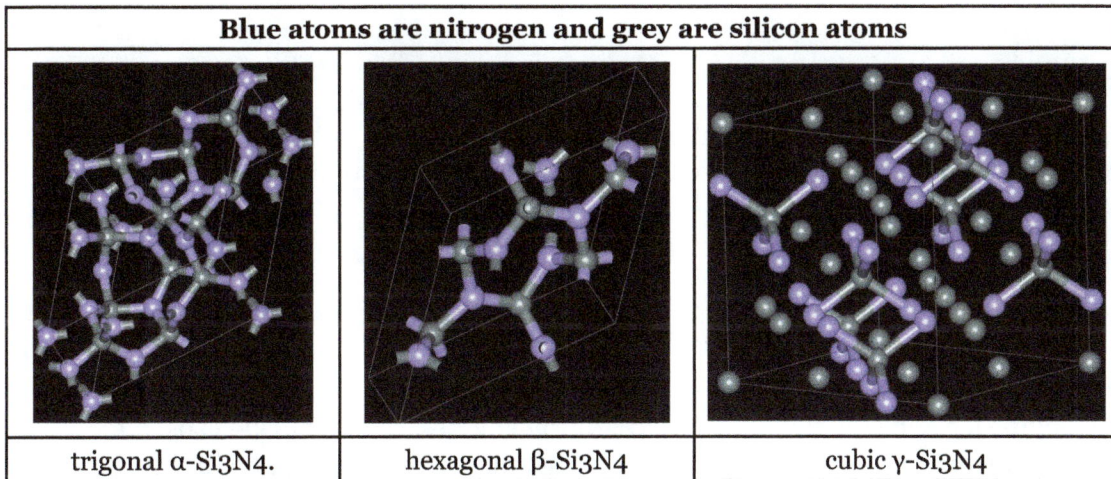

Blue atoms are nitrogen and grey are silicon atoms		
trigonal α-Si3N4.	hexagonal β-Si3N4	cubic γ-Si3N4

There exist three crystallographic structures of silicon nitride (Si3N4), designated as α, β and γ phases. The α and β phases are the most common forms of Si3N4, and can be produced under normal pressure condition. The γ phase can only be synthesized under high pressures and temperatures and has a hardness of 35 GPa.

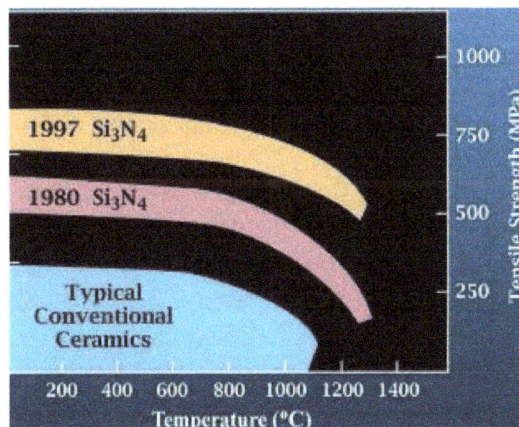

The α- and β-Si3N4 have trigonal (Pearson symbol hP28, space group P31c, No. 159) and hexagonal (hP14, $P6_3$, No. 173) structures, respectively, which are built up by corner-sharing SiN4 tetrahedra. They can be regarded as consisting of layers of silicon and nitrogen atoms in the sequence ABAB... or ABCDABCD... in β-Si3N4 and α-Si3N4, respectively. The AB layer is the same in the α and β phases, and the CD layer in the α phase is related to AB by a c-glide plane. The Si3N4 tetrahedra in β-Si3N4 are interconnected in such a way that tunnels are formed, running parallel with the c axis of the unit cell. Due to the c-glide plane that relates AB to CD, the α structure contains cavities instead of tunnels. The cubic γ-Si3N4 is often designated as c modification in the literature, in analogy with the cubic modification of boron nitride (c-BN). It has a spinel-type structure in which two silicon atoms each coordinate six nitrogen atoms octahedrally, and one silicon atom coordinates four nitrogen atoms tetrahedrally.

The longer stacking sequence results in the α-phase having higher hardness than the β-phase. However, the α-phase is chemically unstable compared with the β-phase. At high temperatures when a liquid phase is present, the α-phase always transforms into the β-phase. Therefore, β-Si3N4 is the major form used in Si3N4 ceramics.

Applications

In general, the main issue with applications of silicon nitride has not been technical performance, but cost. As the cost has come down, the number of production applications is accelerating.

Automobile Industry

One of the major applications of sintered silicon nitride is in automobile industry as a material for engine parts. Those include, in diesel engines, glowplugs for faster start-up; precombustion chambers (swirl chambers) for lower emissions, faster start-up and lower noise; turbocharger for reduced engine lag and emissions. In spark-ignition engines, silicon nitride is used for rocker arm pads for lower wear, turbocharger for lower inertia and less engine lag, and in exhaust gas control valves for increased acceleration. As examples of production levels, there is an estimated more than 300,000 sintered silicon nitride turbochargers made annually.

Bearings

Silicon nitride bearings are both full ceramic bearings and ceramic hybrid bearings with balls in ceramics and races in steel. Silicon nitride ceramics have good shock resistance compared to other ceramics. Therefore, ball bearings made of silicon nitride ceramic are used in performance bearings. A representative example is use of silicon nitride bearings in the main engines of the NASA's Space Shuttle.

Since silicon nitride ball bearings are harder than metal, this reduces contact with the bearing track. This results in 80% less friction, 3 to 10 times longer lifetime, 80% higher speed, 60% less weight, the ability to operate with lubrication starvation, higher corrosion resistance and higher operation temperature, as compared to traditional metal bearings. Silicon nitride balls weigh 79% less than tungsten carbide balls. Silicon nitride ball bearings can be found in high end automotive bearings, industrial bearings, wind turbines, motorsports, bicycles, rollerblades and skateboards. Silicon nitride bearings are especially useful in applications where corrosion, electric or magnetic fields prohibit the use of metals. For example, in tidal flow meters, where seawater attack is a problem, or in electric field seekers.

Si_3N_4 was first demonstrated as a superior bearing in 1972 but did not reach production until nearly 1990 because of challenges associated with reducing the cost. Since 1990, the cost has been reduced substantially as production volume has increased. Although Si3N4 bearings are still 2–5 times more expensive than the best steel bearings, their superior performance and life are justifying rapid adoption. Around 15–20 million Si3N4 bearing balls were produced in the U.S. in 1996 for machine tools and many other applications. Growth is estimated at 40% per year, but could be even higher if ceramic bearings are selected for consumer applications such as in-line skates and computer disk drives.

High-temperature Material

Silicon nitride has long been used in high-temperature applications. In particular, it was identified as one of the few monolithic ceramic materials capable of surviving the severe thermal shock and thermal gradients generated in hydrogen/oxygen rocket engines. To demonstrate this capability in a complex configuration, NASA scientists used advanced rapid prototyping technology to fabricate a one-inch-diameter, single-piece combustion chamber/nozzle (thruster) component. The thruster was hot-fire tested with hydrogen/oxygen propellant and survived five cycles including a 5-minute cycle to a 1320 °C material temperature.

In 2010 silicon nitride was used as the main material in the thrusters of the JAXA space probe Akatsuki.

Medical

Silicon nitride has many orthopedic applications. The material is also an alternative to PEEK (polyether ether ketone) and titanium, which are used for spinal fusion devices. It is silicon nitride's hydrophilic, microtextured surface that contributes to the materials strength, durability and reliability compared to PEEK and titanium.

Metal Working and Cutting Tools

The first major application of Si3N4 was abrasive and cutting tools. Bulk, monolithic silicon nitride is used as a material for cutting tools, due to its hardness, thermal stability, and resistance to wear. It is especially recommended for high speed machining of cast iron. Hot hardness, fracture toughness and thermal shock resistance mean that sintered silicon nitride can cut cast iron, hard steel and nickel based alloys with surface speeds up to 25 times quicker than those obtained with conventional materials such as tungsten carbide. The use of Si 3N4 cutting tools has had a dramatic effect on manufacturing output. For example, face milling of gray cast iron with silicon nitride inserts doubled the cutting speed, increased tool life from one part to six parts per edge, and reduced the average cost of inserts by 50%, as compared to traditional tungsten carbide tools.

Electronics

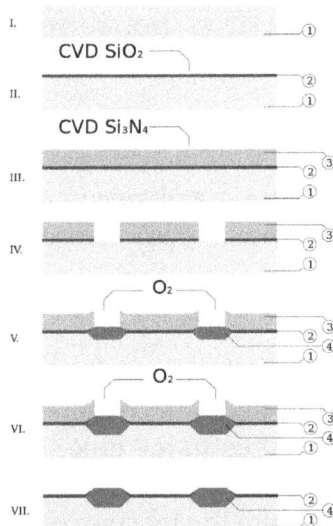

Example of local silicon oxidation through a Si_3N_4 mask

Silicon nitride is often used as an insulator and chemical barrier in manufacturing integrated circuits, to electrically isolate different structures or as an etch mask in bulk micromachining. As a passivation layer for microchips, it is superior to silicon dioxide, as it is a significantly better diffusion barrier against water molecules and sodium ions, two major sources of corrosion and instability in microelectronics. It is also used as a dielectric between polysilicon layers in capacitors in analog chips.

Si_3N_4 cantilever used in atomic force microscopes

Silicon nitride deposited by LPCVD contains up to 8% hydrogen. It also experiences strong tensile stress, which may crack films thicker than 200 nm. However, it has higher resistivity and dielectric strength than most insulators commonly available in microfabrication (10^{16} $\Omega \cdot$cm and 10 MV/cm, respectively).

Not only silicon nitride, but also various ternary compounds of silicon, nitrogen and hydrogen (SiN_xH_y) are used insulating layers. They are plasma deposited using the following reactions:

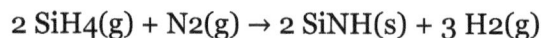

$$2 \, SiH_4(g) + N_2(g) \rightarrow 2 \, SiNH(s) + 3 \, H_2(g)$$

$$SiH_4(g) + NH_3(g) \rightarrow SiNH(s) + 3 \, H_2(g)$$

These SiNH films have much less tensile stress, but worse electrical properties (resistivity 10^6 to 10^{15} $\Omega \cdot$cm, and dielectric strength 1 to 5 MV/cm).

Silicon nitride is also used in xerographic process as one of the layer of the photo drum. Silicon nitride is also used as an ignition source for domestic gas appliances. Because of its good elastic properties, silicon nitride, along with silicon and silicon oxide, is the most popular material for cantilevers — the sensing elements of atomic force microscopes.

History

The first reported preparation was in 1857 by Henri Etienne Sainte-Claire Deville and Friedrich Wöhler. In their method, silicon was heated in a crucible placed inside another crucible packed with carbon to reduce permeation of oxygen to the inner crucible. They reported a product they termed silicon nitride but without specifying its chemical composition. Paul Schuetzenberger first reported a product with the composition of the tetranitride, Si3N4, in 1879 that was obtained by heating silicon with brasque (a paste made by mixing charcoal, coal, or coke with clay which is then used to line crucibles) in a blast furnace. In 1910, Ludwig Weiss and Theodor Engelhardt heated silicon under pure nitrogen to produce Si3N4. E. Friederich and L. Sittig made Si_3N_4 in 1925 via carbothermal reduction under nitrogen, that is, by heating silica, carbon, and nitrogen at 1250–1300 °C.

Silicon nitride remained merely a chemical curiosity for decades before it was used in commercial applications. From 1948 to 1952, the Carborundum Company, Niagara Falls, New York, applied for several patents on the manufacture and application of silicon nitride. By 1958 Haynes (Union Carbide) silicon nitride was in commercial production for thermocouple tubes, rocket nozzles, and boats and crucibles for melting metals. British work on silicon nitride, started in 1953, was aimed at high-temperature parts of gas turbines and resulted in the development of reaction-bonded silicon nitride and hot-pressed silicon nitride. In 1971, the Advanced Research Project Agency of the US Department of Defense placed a US$17 million contract with Ford and Westinghouse for two ceramic gas turbines.

Even though the properties of silicon nitride were well known, its natural occurrence was discovered only in the 1990s, as tiny inclusions (about 2 µm × 0.5 µm in size) in meteorites. The mineral was named nierite after a pioneer of mass spectrometry, Alfred O. C. Nier. This mineral might have been detected earlier, again exclusively in meteorites, by Soviet geologists.

Nanoceramic

Nanoceramic is a type of nanoparticle that is composed of ceramics, which are generally classified as inorganic, heat-resistant, nonmetallic solids made of both metallic and nonmetallic compounds. The material offers unique properties. Macroscale ceramics are brittle and rigid and break upon impact. However, nanoceramics take on a larger variety of functions, including dielectric, ferroelectric, piezoelectric, pyroelectric, ferromagnetic, magnetoresistive, superconductive and electro-optical.

Nanoceramics were discovered in the early 1980s. They were formed using a process called sol-gel which mixes nanoparticles within a solution and gel to form the nanoparticle. Later methods involved sintering (pressure and heat). The material is so small that it has basically no flaws. Larger scale materials have flaws that render them brittle.

In 2014 researchers announced a lasering process involving polymers and ceramic particles to form a nanotruss. This structure was able to recover its original form after repeated crushing.

Properties

Nanoceramics have unique properties because of their size and molecular structure. These properties are often shown in terms of various electrical and magnetic physics phenomenons which include:

- Dielectric - An electrical insulator that can be polarized (having electrons aligned so that there is a negative and positive side of the compound) by an electric field to shorten the distance of electron transfer in an electric current

- Ferroelectric - Dielectric materials that polarize in more than one direction (the negative and positive sides can be flipped via an electric field)

- Piezoelectric - Materials that accumulate an electrical charge under mechanical stress

- Pyroelectric - Material that can produce a temporary voltage given a temperature change

- Ferromagnetic - Materials that can to sustain a magnetic field after magnetization

- Magnetoresistive - Materials that change electrical resistance under an external magnetic field

- Superconductive - Materials that exhibit zero electric resistance when cooled to a critical temperature

- Electro-optical - Materials that change optical properties under an electric field

Nanotruss

Nanoceramic is more than 85% air and is very light, strong, flexible and durable. The fractal nano-truss is a nanostructure architecture made of alumina, or aluminum oxide. Its maximum compression is about 1 micron from a thickness of 50 nanometers. After its compression, it can revert to its original shape without any structural damage.

Synthesis

Sol-gel

One process for making nanoceramics varies is the sol-gel process, also known as chemical solution deposition. This involves a chemical solution, or the sol, made of nanoparticles in liquid phase and a precursor, usually a gel or polymer, made of molecules immersed in a solvent. The sol and gel are mixed to produce an oxide material which are generally a type of ceramic. The excess products (a liquid solvent) are evaporated. The particles desires are then heated in a process called densification to produce a solid product. This method could also be applied to produce a nanocomposite by heating the gel on a thin film to form a nanoceramic layer on top of the film.

Two-photon Lithography

This process uses a laser technique called two-photon lithography to etch out a polymer into a three-dimensional structure. The laser hardens the spots that it touches and leaves the rest un-hardened. The unhardened material is then dissolved to produce a "shell". The shell is then coated with ceramic, metals, metallic glass, etc. In the finished state, the nanotruss of ceramic can be flattened and revert to its original state.

Sintering

In another approach sintering was used to consolidate nanoceramic powders using high temperatures. This resulted in a rough material that damages the properties of ceramics and requires more time to obtain an end product. This technique also limits the possible final geometries. Microwave sintering was developed to overcome such problems. Radiation is produced from a magnetron, which produces electromagnetic waves to vibrate and heat the powder. This method allows for heat to be instantly transferred across the entire volume of material instead of from the outside in.

The nanopowder is placed in an insulation box composed of low insulation boards to allow the microwaves to pass through it.The box increases temperature to aid absorption. Inside the boxes are suspectors that absorb microwaves at room temperature to initialize the sintering process. The microwave heats the suspectors to about 600 °C, sufficient to trigger the nanoceramics to absorb the microwaves.

History

In the early 1980s, the first nanoparticles, specifically nanoceramics were formed, using sol-gel. This process was replaced by sintering in the early 2000s and then by microwave sintering. None of these techniques proved suitable for large scale production.

In 2002, researchers tried to reverse engineer the microstructure of seashells to strengthen ceramics. They discovered that seashells' durability come from their "microarchitecture". Research began to focus on how ceramics could employ such an architecture.

In 2012 researchers replicated the sea sponge's structure using ceramics and the nanoarchitecture called nanotruss. As of 2015 the largest result is a 1mm cube. The lattice structure compresses up to 85% of its original thickness and can recover to its original form. These lattices are stabilized into triangles with cross-members for structural integrity and flexibility.

Applications

Medical technology used nanoceramics for bone repair. It has been suggested for areas including energy supply and storage, communication, transportation systems, construction and medical technology. Their electrical properties may allow energy to be transferred efficiencies approaching 100%. Nanotrusses may be eventually applicable for building materials, replacing concrete or steel.

Coade Stone

Father Thames, a Coade stone sculpture by John Bacon, in the grounds of Ham House

Coade stone or *Lithodipyra* was stoneware that was often described as an artificial stone in the late 18th and early 19th centuries. It was used for moulding Neoclassical statues, architectural decorations and garden ornaments that were both of the highest quality and remain virtually weatherproof today. Produced by appointment to George III and the Prince Regent, it features on St George's Chapel, Windsor; The Royal Pavilion, Brighton; Carlton House, London; the Royal Naval College, Greenwich; and a large quantity was used in the refurbishment of Buckingham Palace in the 1820s.

Lithodipyra was first created around 1770 by Eleanor Coade who ran *Coade's Artificial Stone Manufactory*, *Coade and Sealy*, and *Coade* in Lambeth, London, from 1769 until her death in 1821, after which *Lithodipyra* continued to be manufactured by her last business partner William Croggon until 1833.

The recipe and techniques for producing Coade stone have been rediscovered by the team at Coade ltd., which now reproduce a range of Coade sculpture at their workshops in Wilton.

History

Lion Gate, an entrance into Kew Gardens, with its Coade stone lion statue on top.

In 1769 Mrs Coade bought Daniel Pincot's struggling artificial stone business at Kings Arms Stairs, Narrow Wall, Lambeth, a site now under the Royal Festival Hall. This business developed into *Coade's Artificial Stone Manufactory* with Eleanor in charge, such that within two years (1771) she fired Pincot for 'representing himself as the chief proprietor'.

Mrs Coade did not invent 'artificial stone' - various inferior quality precursors having been both patented and manufactured over the previous forty (or sixty) years - but she was probably responsible for perfecting both the clay recipe and the firing process. It is possible that Pincot's business was a continuation of that run nearby by Richard Holt, who had taken out two patents in 1722 for a kind of liquid metal or stone and another for making china without the use of clay, but there were many start-up 'artificial stone' businesses in the early 18th century of which only Mrs Coade's succeeded.

The company did well, and boasted an illustrious list of customers such as George III and members of the English nobility. In 1799 Mrs Coade appointed her cousin John Sealy (her mother's sister Mary's son), already working as a modeller, as a partner in her business, which then traded as 'Coade and Sealy' until his death in 1813 when it reverted to just 'Coade'.

In 1799 she opened a show room *Coade's Gallery* on *Pedlar's Acre* at the Surrey end of Westminster Bridge Road to display her products.

In 1813 Mrs Coade took on William Croggan from Grampound in Cornwall, a sculptor and distant relative by marriage (second cousin once removed). He managed the factory until her death eight years later in 1821 whereby he bought the factory from the executors for c. £4000. Croggan supplied a lot of Coade stone for Buckingham Palace; however, he went bankrupt in 1833 and died two years later. Trade declined, and production came to an end in the early 1840s.

In 2000 Coade ltd started reproducing Coade stone statues also creating new sculptures and architectural ornament, using the original recipes and methods of the eighteenth century.

The Material

Home of Eleanor Coade, Belmont House, in Lyme Regis, Dorset, with Coade stone ornamental façade

Its colours varied from light grey to light yellow (or even beige) and its surface is best described as having a matte finish.

The ease with which the product could be moulded into complex shapes made it ideal for large statues, sculptures and sculptural façades. Moulds were often kept for many years, for repeated use. One-offs were clearly much more expensive to produce, as they had to carry the entire cost of creating the mould.

One of the more striking features of Coade stone is its incredible resistance to weathering, often faring better than most types of stone in London's harsh environment. Examples of Coade stone-work have survived very well; prominent examples are listed below, having survived without apparent wear and tear for 150 years.

As a material, Coade stone was replaced by Portland cement as a form of artificial stone and it appears to have been largely phased out by the 1840s.

Quality Controversy

Although Coade stone's reputation for both weather resistance and manufacturing quality is virtually untarnished, three sources describe Rossi's statue of George IV erected in the Royal Crescent, Brighton as "unable to withstand the weathering effects of sea-spray and strong wind: such that, by 1807 the fingers on the sculpture's left hand had been destroyed, and soon afterwards the whole right arm dropped off." By contrast however *Fashionable Brighton, 1820-1860* by Antony Dale (online) describes similar damage as 'wore badly' but does not attribute 'broken fingers, nose, mantle and arm on an unloved statue' to weathering or poor quality Coade stone. In 1819, after considerable complaints, the relic was removed and its present state is undocumented. A few works produced by Coade, mainly dating from the later period, have shown poor resistance to weathering due to a bad firing in the kiln where the material was not brought up to a sufficient temperature.

The Formula

Contrary to popular belief, the recipe for Coade stone still exists, and can be produced by Coade ltd. Rather than being based on cement (as concrete articles are), it is a ceramic material.

Its manufacture required special skills: extremely careful control and skill in kiln firing, over a period of days. This skill is even more remarkable when the potential variability of kiln temperatures at that time is considered. Mrs Coade's factory was the only really successful manufacturer.

The formula used was:

- 10% of grog
- 5-10% of crushed flint
- 5-10% fine quartz
- 10% crushed soda lime glass.
- 60-70% Ball clay from Dorset and Devon

This mixture was also referred to as "fortified clay" which was then inserted after kneading into a kiln which would fire the material at a temperature of 1,100 °C for over four days.

A number of different variations of the recipe were used, depending on the size and fineness of detail in the work a different size and proportion of grog was used. In many pieces a combination of fine grogged Coade clay was used on the surface for detail, backed up by a more heavily grogged mixture for strength.

Examples

Over 650 pieces are still in existence worldwide.

The Red Lion, aka the South Bank Lion, on Westminster Bridge. Modelled by William F. Woodington and Grade II* listed by English Heritage

Schomberg House circa 1850

Captain William Bligh's Tomb surmounted by a breadfruit in a bowl

London Lodge (1793), Highclere Castle, Hampshire. Brick but Coade stone dressed, and wings (1840), Highclere Castle, Hampshire, May 2014.

- Mrs Coade's country home, Belmont House in Lyme Regis, Dorset, displays examples of Coade stone on its façade.

- The South Bank Lion at the south end of Westminster Bridge in central London originally stood atop the old Red Lion Brewery, on the Lambeth bank of the River Thames. When the brewery was demolished in 1950, to make way for the South Bank Site of the 1951 Festival of Britain, the Lion was taken down and moved to Station Approach Waterloo, painted red as the symbol of British Rail on high plinth. When removed, the initials of the sculptor William F. Woodington and the date, 24 May 1837, were discovered under one of its paws. The fine details still remain clear after 170 years of London's corrosive atmosphere, caused by heavy use of coal throughout the 19th and first half of the 20th centuries. The red paint was removed to reveal the fine Coade stone surface to view. In 1966, the statue was moved from outside Waterloo station to its current location.

- Duff House Mausoleum, Wrack Woods, Banff, Aberdeenshire, Scotland. The second Earl of Fife built this mausoleum for his family tombs in 1791, possibly on the site of a Carmelite friary. Built before the Gothic Revival, this is an example of "Gothick" architecture. Typically of the Georgians the carvings, including the monument to the first Earl, are in ceramic Coade stone.

- Nelson's Memorial at Burnham Thorpe

- St Botolph-without-Bishopsgate Church Hall, London, pair of statues of schoolchildren on the front of this former School House, replicas outside, listed originals now inside the Hall.

- The statue and ornaments on Nelson's Column, Montreal, built 1809

- Britannia Monument in Great Yarmouth

- The Lion Gate, Mote Park, Roscommon, Ireland, built 1787

- Nelson's Pediment on the Old Royal Naval College, Greenwich, regarded by the Coade workers as the finest of all their work.

- Twinings' first ever (and still operating) shop's frontispiece, in the Strand, London opposite the Royal Courts of Justice, rediscovered under soot after a century.

- Schomberg House on Pall Mall, London

- Captain Bligh's tomb (in the churchyard of St Mary's Lambeth)

- Lord Hill's Column, Shrewsbury, Shropshire

- Rio de Janeiro's zoo entrance

- St Mary's Church gate, Tremadog, Gwynedd, Wales.

- Richmond upon Thames. Two examples of the River God, one outside Ham House, the other in Terrace Gardens.

- Buckingham Palace (in a section not open to the public)

- Castle Howard

- A couple of large ornate urns in the Italian Garden at Chiswick House, London

- Royal Pavilion in Brighton

- Imperial War Museum (sculptural reliefs above the entrance)

- The Buttermarket in Chichester, which was designed by John Nash (coat of arms engraved with "Coade & Sealey 1808")

- Saxham Hall, Suffolk has an Umbrello (shelter) constructed of Coade stone in the grounds

- The lion and unicorn statues over their respective gates into Kew Gardens.

- Burton Constable Hall in Holderness, East Riding of Yorkshire displays 3 figures and a number of 'medallions' above the doors and windows of the Orangerie.

- The Royal Arms of Queen Charlotte above 8 Argyle Street, Bath.

- The keystone, featuring a carving of the head of Silenus, above the entry to The Rossborough Inn, a historic building at the University of Maryland, College Park, in the United States.

- Edinburgh, three Coade Stone columns on Portobello Beach

Ceramic Matrix Composite

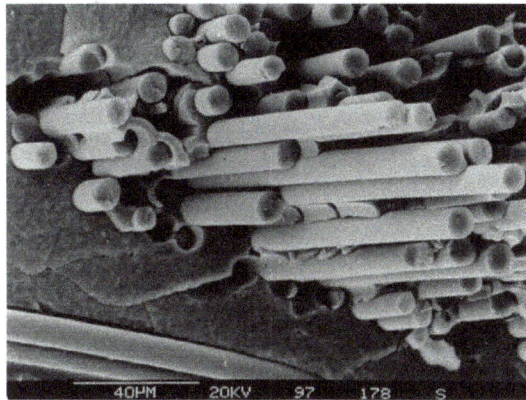

Fracture surface of a fiber-reinforced ceramic composed of SiC fibers and SiC matrix. The fiber pull-out mechanism shown is the key to CMC properties.

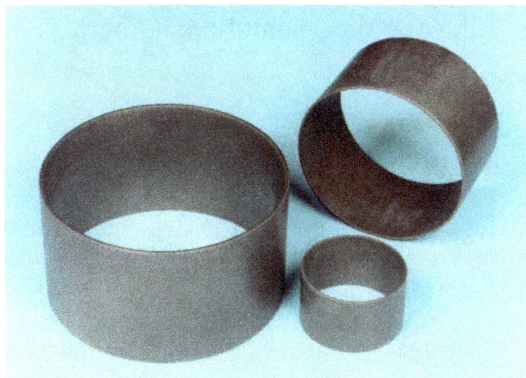

CMC shaft sleeves with outer diameters between 100 and 300 mm for ceramic slide bearings of pumps.

Ceramic matrix composites (CMCs) are a subgroup of composite materials as well as a subgroup of technical ceramics. They consist of ceramic fibres embedded in a ceramic matrix, thus forming a ceramic fibre reinforced ceramic (CFRC) material. The matrix and fibres can consist of any ceramic material, whereby carbon and carbon fibres can also be considered a ceramic material.

Introduction

The motivation to develop CMCs was to overcome the problems associated with the conventional technical ceramics like alumina, silicon carbide, aluminium nitride, silicon nitride or zirconia – they fracture easily under mechanical or thermo-mechanical loads because of cracks initiated by small defects or scratches. The crack resistance is – like in glass – very low. To increase the crack resistance or fracture toughness, particles (so-called monocrystalline *whiskers* or *platelets*) were embedded into the matrix. However, the improvement was limited, and the products have found application only in some ceramic cutting tools. So far only the integration of long multi-strand fibres has drastically increased the crack resistance, elongation and thermal shock resistance, and resulted in several new applications.

Carbon (C), special silicon carbide (SiC), alumina (Al_2O_3) and mullite (Al_2O_3–SiO_2) fibres are most commonly used for CMCs. The matrix materials are usually the same, that is C, SiC, alumina and mullite.

Generally, CMC names include a combination of *type of fibre/type of matrix*. For example, *C/C* stands for carbon-fibre-reinforced carbon (carbon/carbon), or *C/SiC* for carbon-fibre-reinforced silicon carbide. Sometimes the manufacturing process is included, and a C/SiC composite manufactured with the liquid polymer infiltration (LPI) process is abbreviated as *LPI-C/SiC*.

The important commercially available CMCs are C/C, C/SiC, SiC/SiC and Al_2O_3/Al_2O_3. They differ from conventional ceramics in the following properties, presented in more detail below:

- Elongation to rupture up to 1%

- Strongly increased fracture toughness

- Extreme thermal shock resistance

- Improved dynamical load capability

- Anisotropic properties following the orientation of fibers

Manufacture

The manufacturing processes usually consist of the following three steps:

1. Lay-up and fixation of the fibres, shaped as the desired component

2. Infiltration of the matrix material

3. Final machining and, if required, further treatments like coating or impregnation of the intrinsic porosity.

The first and the last step are almost the same for all CMCs: In step one, the fibres, often named rovings, are arranged and fixed using techniques used in fibre-reinforced plastic materials, such as lay-up of fabrics, filament winding, braiding and knotting. The result of this procedure is called *fibre-preform* or simply *preform.*

For the second step, five different procedures are used to fill the ceramic matrix in between the fibres of the preform:

1. Deposition out of a gas mixture

2. Pyrolysis of a pre-ceramic polymer

3. Chemical reaction of elements

4. Sintering at a relatively low temperature in the range 1000–1200 °C

5. Electrophoretic deposition of a ceramic powder

Procedures one, two and three find applications with non-oxide CMCs, whereas the fourth one is used for oxide CMCs; combinations of these procedures are also practised. The fifth procedure is not yet established in industrial processes. All procedures have sub-variations, which differ in technical details. All procedures yield a porous material.

The third and final step of machining – grinding, drilling, lapping or milling – has to be done with diamond tools. CMCs can also be processed with a water jet, laser, or ultrasonic machining.

Ceramic Fibres

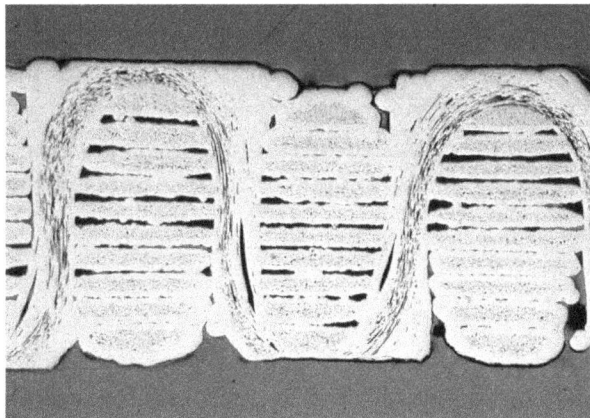

Micrograph of a SiC/SiC ceramic composite with a woven three-dimensional fibre structure

Ceramic fibres in CMCs can have a polycrystalline structure, as in conventional ceramics. They can also be amorphous or have inhomogeneous chemical composition, which develops upon pyrolysis of organic precursors. The high process temperatures required for making CMCs preclude the use of organic, metallic or glass fibres. Only fibres stable at temperatures above 1000 °C can be used, such as fibres of alumina, mullite, SiC, zirconia or carbon. Amorphous SiC fibres have an elongation capability above 2% – much larger than in conventional ceramic materials (0.05 to 0.10%). The reason for this property of SiC fibres is that most of them contain additional elements like oxygen, titanium and/or aluminium yielding a tensile strength above 3 GPa. These enhanced elastic properties are required for various three-dimensional fibre arrangements in textile fabrication, where a small bending radius is essential.

Manufacturing Procedures

Matrix Deposition from a Gas Phase

Chemical vapour deposition (CVD) is well suited for this purpose. In the presence of a fibre pre-form, CVD takes place in between the fibres and their individual filaments and therefore is called chemical vapour infiltration (CVI). One example is the manufacture of C/C composites: a C-fibre preform is exposed to a mixture of argon and a hydrocarbon gas (methane, propane, etc.) at a pressure of around or below 100 kPa and a temperature above 1000 °C. The gas decomposes depositing carbon on and between the fibers. Another example is the deposition of silicon carbide, which is usually conducted from a mixture of hydrogen and methyl-trichlorosilane (MTS, CH_3SiCl_3; it is also common in silicone production). Under defined condition this gas mixture deposits fine and crystalline silicon carbide on the hot surface within the preform. This CVI procedure leaves a body with a porosity of about 10–15%, as access of reactants to the interior of the preform is increasingly blocked by deposition on the exterior.

Matrix Forming via Pyrolysis of C- and Si-containing Polymers

Hydrocarbon polymers shrink during pyrolysis, and upon outgassing form carbon with an amorphous, glass-like structure, which by additional heat treatment can be changed to a more graphite-like structure. Other special polymers, where some carbon atoms are replaced by silicon atoms, the so-called polycarbosilanes, yield amorphous silicon carbide of more or less stoichiometric composition. A large variety of such SiC-, SiNC-, or SiBNC-producing precursors already exist and more are being developed. To manufacture a CMC material, the fibre preform is infiltrated with the chosen polymer. Subsequent curing and pyrolysis yield a highly porous matrix, which is undesirable for most applications. Further cycles of polymer infiltration and pyrolysis are performed until the final and desired quality is achieved. Usually five to eight cycles are necessary. The process is called *liquid polymer infiltration* (LPI), or *polymer infiltration and pyrolysis* (PIP). Here also a porosity of about 15% is common due to the shrinking of the polymer. The porosity is reduced after every cycle.

Matrix Forming via Chemical Reaction

With this method, one material located between the fibres reacts with a second material to form the ceramic matrix. Some conventional ceramics are also manufactured by chemical reactions. For example, reaction-bonded silicon nitride (RBSN) is produced through the reaction of silicon powder with nitrogen, and porous carbon reacts with silicon to form reaction bonded silicon carbide, a silicon carbide which contains inclusions of a silicon phase. An example of CMC manufacture, which was introduced for the production of ceramic brake discs, is the reaction of silicon with a porous preform of C/C. The process temperature is above 1414 °C, that is above the melting point of silicon, and the process conditions are controlled such that the carbon fibres of the C/C-preform almost completely retain their mechanical properties. This process is called *liquid silicon infiltration* (LSI). Sometimes, and because of its starting point with C/C, the material is abbreviated as *C/C-SiC*. The material produced in this process has a very low porosity of about 3%.

Matrix Forming via Sintering

This process is used to manufacture oxide fibre/oxide matrix CMC materials. Since most ceramic fibres can not withstand the normal sintering temperatures of above 1600 °C, special precursor liquids are used to infiltrate the preform of oxide fibres. These precursors allow sintering, that is ceramic-forming processes, at temperatures of 1000–1200 °C. They are, for example, based on mixtures of alumina powder with the liquids tetra-ethyl-orthosilicate (as Si donor) and aluminium-butylate (as Al donor), which yield a mullite matrix. Other techniques, such as sol-gel chemistry, are also used. CMCs obtained with this process usually have a high porosity of about 20%.

Matrix formed via electrophoresis

In the electrophoretic process, electrically charged particles dispersed in a special liquid are transported through an electric field into the preform, which has the opposite electrical charge polarity. This process is under development, and is not yet used industrially. Some remaining porosity must be expected here, too.

Properties

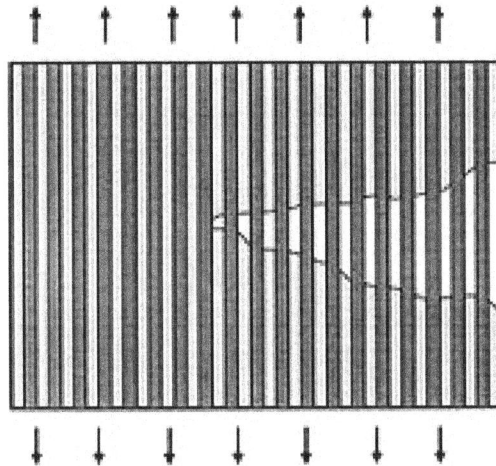

Scheme of crack bridges at the crack tip of ceramic composites.

Mechanical Properties

Basic Mechanism of Mechanical Properties

The high fracture toughness or crack resistance mentioned above is a result of the following mechanism: under load the ceramic matrix cracks, like any ceramic material, at an elongation of about 0.05%. In CMCs the embedded fibres bridge these cracks. This mechanism works only when the matrix can slide along the fibres, which means that there must be a weak bond between the fibres and matrix. A strong bond would require a very high elongation capability of the fibre bridging the crack, and would result in a brittle fracture, as with conventional ceramics. The production of CMC material with high crack resistance requires a step to weaken this bond between the fibres and matrix. This is achieved by depositing a thin layer of pyrolytic carbon or boron nitride on the fibres, which weakens the bond at the fibre/matrix interface (sometimes "interface"), leading to the fibre pull-out at crack surfaces, as shown in the SEM picture at the top of this article. In oxide-CMCs, the high porosity of the matrix is sufficient to establish the weak bond.

Properties Under Tensile and Bending Loads, Crack Resistance

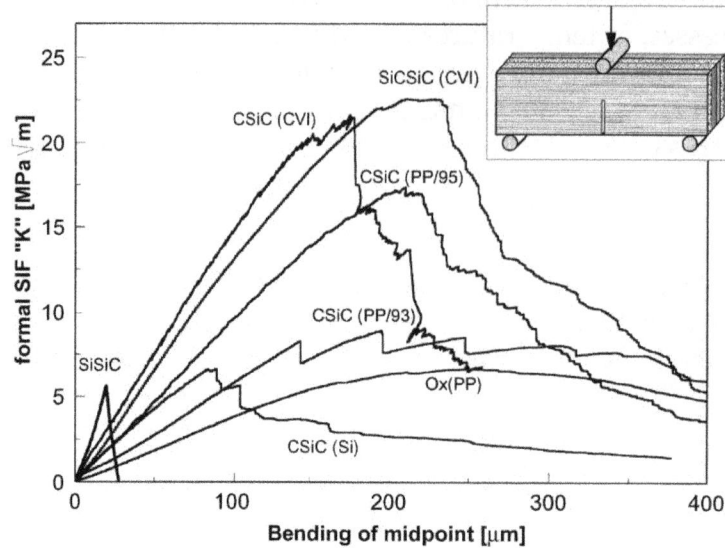

Curves of toughness measurements of various ceramic composites and SiSiC
Legend: SiSiC: conventional SiSiC, SiCSiC(CVI) and CSiC(CVI): SiC/SiC and C/SiC manufactured in CVI processes, CSiC(95) und CSiC(93): C/SiC manufactured by the LPI-method, Ox(PP): oxide ceramic composite, CSiC(Si): C/SiC manufactured via the LSI process.

The influence and quality of the fibre interface can be evaluated through mechanical properties. Measurements of the crack resistance were performed with notched specimens in so-called single-edge-notch-bend (SENB) tests. In fracture mechanics, the measured data (force, geometry and crack surface) are normalized to yield the so-called stress intensity factor (SIF), K_{Ic}. Because of the complex crack surface the real crack surface area can not be determined for CMC materials. The measurements therefore use the initial notch as the crack surface, yielding the *formal SIF* shown in the figure. This requires identical geometry for comparing different samples. The area under these curves thus gives a relative indication of the energy required to drive the crack tip through the sample (force times path length gives energy). The maxima indicate the load level necessary to propagate the crack through the sample. Compared to the sample of conventional SiSiC ceramic, two observations can be made:

- All tested CMC materials need up to several orders of magnitude more energy to propagate the crack through the material.

- The force required for crack propagation varies between different types of CMCs.

Type of material	Al_2O_3/Al_2O_3	Al_2O_3	CVI-C/SiC	LPI-C/SiC	LSI-C/SiC	SiSiC
Porosity (%)	35	<1	12	12	3	<1
Density (g/cm³)	2.1	3.9	2.1	1.9	1.9	3.1
Tensile strength (MPa)	65	250	310	250	190	200
Elongation (%)	0.12	0.1	0.75	0.5	0.35	0.05
Young's modulus (GPa)	50	400	95	65	60	395
Flexural strength (MPa)	80	450	475	500	300	400

In the table, CVI, LPI, and LSI denote the manufacturing process of the C/SiC-material. Data of the oxide CMC and SiSiC are taken from manufacturer data sheets. Tensile strength of SiSiC and Al_2O_3 were calculated from measurements of elongation to fracture and Young's modulus, since generally only bending strength data are available for those ceramics. Averaged values are given in the table, and significant differences, even within one manufacturing route, are possible.

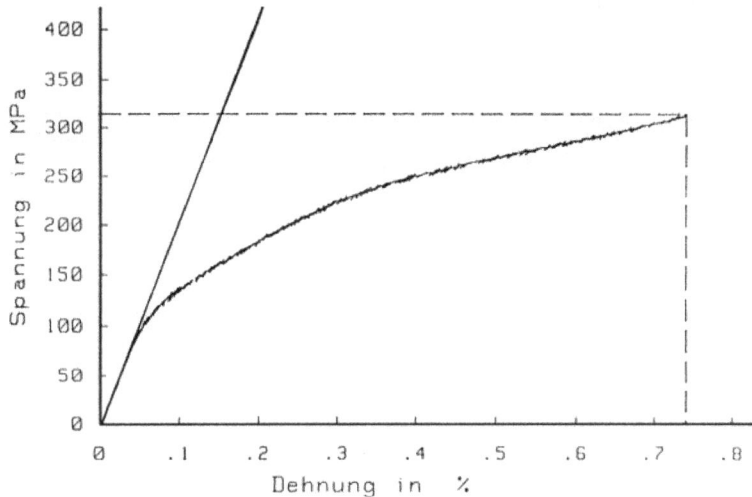

Stress-strain curve of a tensile test for CVI-SiC/SiC.

Tensile tests of CMCs usually show nonlinear stress-strain curves, which look as if the material deforms plastically. It is called *quasi-plastic*, because the effect is caused by the microcracks, which are formed and bridged with increasing load. Since the Young's modulus of the load-carrying fibres is generally lower than that of the matrix, the slope of the curve decreases with increasing load.

Curves from bending tests look similar to those of the crack resistance measurements shown above.

The following features are essential in evaluating bending and tensile data of CMCs:

- CMC materials with a low matrix content (down to zero) have a high tensile strength (close to the tensile strength of the fibre), but low bending strength.

- CMC materials with a low fibre content (down to zero) have a high bending strength (close to the strength of the monolithic ceramic), but no elongation beyond 0.05% under tensile load.

The primary quality criterion for CMCs is the crack resistance behaviour or fracture toughness.

Other Mechanical Properties

In many CMC components the fibres are arranged as 2-dimensional (2D) stacked plain or satin weave fabrics. Thus the resulting material is anisotropic or, more specifically, orthotropic. A crack between the layers is not bridged by fibres. Therefore, the interlaminar shear strength (ILS) and the strength perpendicular to the 2D fiber orientation are low for these materials. Delamination can occur easily under certain mechanical loads. Three-dimensional fibre structures can improve this situation.

Material	CVI-C/ SiC	LPI-C/ SiC	LSI-C/ SiC	CVI-SiC/ SiC
Interlaminar shear strength (MPa)	45	30	33	50
Tensile strength vertical to fabric plane (MPa)	6	4	–	7
Compressive strength vertical to fabric plane (MPa)	500	450	–	500

The compressive strengths shown in the table are lower than those of conventional ceramics, where values above 2000 MPa are common; this is a result of porosity.

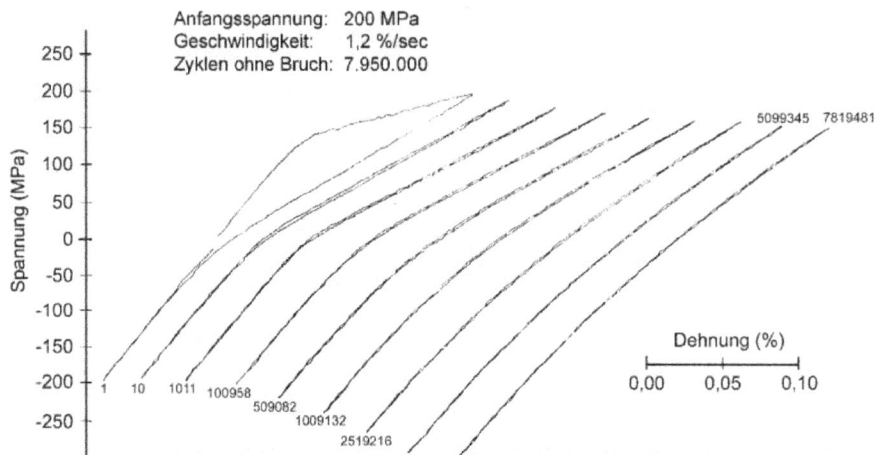

Strain controlled LCF-test for a CVI-SiC/SiC-specimen.

The composite structure allows high dynamical loads. In the so-called low-cycle-fatigue (LCF) or high-cycle-fatigue (HCF) tests the material experiences cyclic loads under tensile and compressive (LCF) or only tensile (HCF) load. The higher the initial stress the shorter the lifetime and the smaller the number of cycles to rupture. With an initial load of 80% of the strength, a SiC/SiC sample survived about 8 million cycles.

The Poisson's ratio shows an anomaly when measured perpendicular to the plane of the fabric, because interlaminar cracks increase the sample thickness.

Thermal and Electrical Properties

The thermal and electrical properties of the composite are a result of its constituents, namely fibres, matrix and pores as well as their composition. The orientation of the fibres yields anisotropic data. Oxide CMCs are very good electrical insulators, and because of their high porosity their thermal insulation is much better than that of conventional oxide ceramics.

The use of carbon fibres increases the electrical conductivity, provided the fibres contact each other and the voltage source. Silicon carbide matrix is a good thermal conductor. Electrically, it is a semiconductor, and its resistance therefore decreases with increasing temperature. Compared to (poly)crystalline SiC, the amorphous SiC fibres are relatively poor conductors of heat and electricity.

Material	CVI-C/ SiC	LPI-C/ SiC	LSI-C/ SiC	CVI-SiC/ SiC	SiSiC
Thermal conductivity (p) [W/(m·K)]	15	11	21	18	>100
Thermal conductivity (v) [W/(m·K)]	7	5	15	10	>100
Linear expansion (p) [10^{-6}·1/K]	1.3	1.2	0	2.3	4
Linear expansion (v) [10^{-6}·1/K]	3	4	3	3	4
Electrical resistivity (p) [Ω·cm]	–	–	–	–	50
Electrical resistivity (v) [Ω·cm]	0.4	–	–	5	50

Comments for the table: (p) and (v) refer to data parallel and vertical to fibre orientation of the 2D-fiber structure, respectively. LSI material has the highest thermal conductivity because of its low porosity – an advantage when using it for brake discs. These data are subject to scatter depending on details of the manufacturing processes.

Conventional ceramics are very sensitive to thermal stress because of their high Young's modulus and low elongation capability. Temperature differences and low thermal conductivity create locally different elongations, which together with the high Young's modulus generate high stress. This results in cracks, rupture and brittle failure. In CMCs, the fibres bridge the cracks, and the components show no macroscopic damage, even if the matrix has cracked locally. The application of CMCs in brake disks demonstrates the effectiveness of ceramic composite materials under extreme thermal shock conditions.

Corrosion Properties

Data on the corrosion behaviour of CMCs are scarce except for oxidation at temperatures above 1000 °C. These properties are determined by the constituents, namely the fibres and matrix. Ceramic materials in general are very stable to corrosion. The broad spectrum of manufacturing techniques with different sintering additives, mixtures, glass phases and porosities are crucial for the results of corrosion tests. Less impurities and exact stoichiometry lead to less corrosion. Amorphous structures and non-ceramic chemicals frequently used as sintering aids are starting points of corrosive attack.

Alumina

Pure alumina shows excellent corrosion resistivity against most chemicals. Amorphous glass and silica phases at the grain boundaries determine the speed of corrosion in concentrated acids and bases and result in creep at high temperatures. These characteristics limit the use of alumina. For molten metals, alumina is used only with gold and platinum.

Alumina fibres

These fibres demonstrate corrosion properties similar to alumina, but commercially available fibres are not very pure and therefore less resistant. Because of creep at temperatures above 1000 °C, there are only few applications for oxide CMCs.

Carbon

The most significant corrosion of carbon occurs in presence of oxygen above about 500 °C. It burns to form carbon dioxide and/or carbon monoxide. It also oxidises in strong oxidizing agents like concentrated nitric acid. In molten metals it dissolves and forms metal carbides. Carbon fibres do not differ from carbon in their corrosion behaviour.

Silicon carbide

Pure silicon carbide is one of the most corrosion-resistant materials. Only strong bases, oxygen above about 800 °C, and molten metals react with it to form carbides and silicides. The reaction with oxygen forms SiO_2 and CO_2, whereby a surface layer of SiO_2 slows down subsequent oxidation (*passive oxidation*). Temperatures above about 1600 °C and a low partial pressure of oxygen result in so-called *active oxidation*, in which CO, CO_2 and gaseous SiO are formed causing rapid loss of SiC. If the SiC matrix is produced other than by CVI, corrosion-resistance is not as good. This is a consequence of porosity in the amorphous LPI, and residual silicon in the LSI-matrix.

Silicon carbide fibres

Silicon carbide fibres are produced via pyrolysis of organic polymers, and therefore their corrosion properties are similar to those of the silicon carbide found in LPI-matrices. These fibres are thus more sensitive to bases and oxidizing media than pure silicon carbide.

Applications

CMC materials overcome the major disadvantages of conventional technical ceramics, namely brittle failure and low fracture toughness, and limited thermal shock resistance. Therefore, their applications are in fields requiring reliability at high-temperatures (beyond the capability of metals) and resistance to corrosion and wear. These include:

- Heat shield systems for space vehicles, which are needed during the re-entry phase, where high temperatures, thermal shock conditions and heavy vibration loads take place.

- Components for high-temperature gas turbines such as combustion chambers, stator vanes and turbine blades.

- Components for burners, flame holders, and hot gas ducts, where the use of oxide CMCs has found its way.

- Brake disks and brake system components, which experience extreme thermal shock (greater than throwing a glowing part of any material into water).

- Components for slide bearings under heavy loads requiring high corrosion and wear resistance.

In addition to the foregoing, CMCs can be used in applications, which employ conventional ceramics or in which metal components have limited lifetimes due to corrosion or high temperatures.

Developments for Applications in Space

During the re-entry phase of space vehicles, the heat shield system is exposed to temperatures above 1500 °C for a few minutes. Only ceramic materials are able to survive such conditions without significant damage, and among ceramics only CMCs can adequately handle thermal shocks. The development of CMC-based heat shield systems promises the following advantages:

- Reduced weight

- Higher load carrying capacity of the system

- Reusability for several re-entries

- Better steering during the re-entry phase with CMC flap systems

NASA-space vehicle X-38 during a test flight

In these applications the high temperatures preclude the use of oxide fibre CMCs, because under the expected loads the creep would be too high. Amorphous silicon carbide fibres lose their strength due to re-crystallization at temperatures above 1250 °C. Therefore, carbon fibres in a silicon carbide matrix (C/SiC) are used in development programs for these applications. The European program HERMES of ESA, started in the 1980s and for financial reasons abandoned in 1992, has produced first results. Several follow-up programs focused on the development, manufacture, and qualification of nose cap, leading edges and steering flaps for the NASA space vehicle X-38.

Pair of steering flaps for the NASA-space vehicle X-38. Size: 1.5×1.5×0.15 m, mass: 68 kg each, various components are mounted using more than 400 CVI-C/SiC screws and nuts.

This development program has qualified the use of C/SiC bolts and nuts, and the bearing system of the flaps. The latter were ground-tested at the DLR in Stuttgart, Germany, under expected conditions of the re-entry phase: 1600 °C, 4 tonnes load, oxygen partial pressure similar to re-entry conditions, and simultaneous bearing movements of four cycles per second. A total of five re-entry phases was simulated. Furthermore, oxidation protection systems were developed and qualified to prevent burnout of the carbon fibres. After mounting of the flaps, mechanical ground tests were performed successfully by NASA in Houston, Texas, US. The next test – a real re-entry of the unmanned vehicle X-38 – was cancelled for financial reasons. One of the space shuttles would have brought the vehicle into orbit, from where it would have returned to the Earth.

These qualifications were promising for only this application. The high-temperature load lasts only around 20 minutes per re-entry, and for reusability, only about 30 cycles would be sufficient. For industrial applications in hot gas environment, though, several hundred cycles of thermal loads and up to many thousands hours of lifetime are required.

The Intermediate Experimental Vehicle (IXV), a project initiated by ESA in 2009, is Europe's first lifting body reentry vehicle. Developed by Thales Alenia Space, the IXV is scheduled to make its first flight in 2014 on the fourth Vega mission (VV04) over the Gulf of Guinea. More than 40 European companies contributed to its construction. The thermal protection system for the underside of the vehicle, comprising the nose, leading edges and lower surface of the wing, were designed and made by Herakles using a ceramic matrix composite (CMC), carbon/silicon-carbide (C/SiC). These components will function as the vehicle's heat shield during its atmospheric reentry.

Developments for Gas Turbine Components

The use of CMCs in gas turbines would permit higher turbine inlet temperatures, which would improve turbine efficiency. Because of the complex shape of stator vanes and turbine blades, the development was first focused on the combustion chamber. In the US, a combustor made of SiC/SiC with a special SiC fiber of enhanced high-temperature stability was successfully tested for 15,000 hours. SiC oxidation was substantially reduced by the use of an oxidation protection coating consisting of several layers of oxides. The engine collaboration between General Electric and Rolls-Royce is studying the use of CMC stator vanes in the hot section of the F136 turbofan engine, an engine which failed to beat the Pratt and Whitney F-135 for use in the Joint Strike Fighter. The engine joint venture, CFM International is also considering the use of CMC parts to reduce weight in its Leap-X demonstrator engine program, which is aimed at providing next-generation turbine engines for narrow-body airliners. CMC parts are also being studied for stationary applications in both the cold and hot sections of the engines, since stresses imposed on rotating parts would require further development effort. Generally, a successful application in turbines still needs a lot of technical and cost reduction work for all high-temperature components to justify the efficiency gain. Furthermore, cost reduction for fibers, manufacturing processes and protective coatings is essential.

Application of Oxide CMC in Burner and Hot Gas Ducts

Oxygen-containing gas at temperatures above 1000 °C is rather corrosive for metal and silicon carbide components. Such components, which are not exposed to high mechanical stress, can be made of oxide CMCs, which can withstand temperatures up to 1200 °C. The gallery below shows the flame holder of a crisp bread bakery as tested after for 15,000 hours, which subsequently operated for a total of more than 20,000 hours.

Oxide CMC flame holder	Ventilator for hot gases	Lifting gate, oxide CMC	Lifting gate in the field

Flaps and ventilators circulating hot, oxygen-containing gases can be fabricated in the same shape as their metal equivalents. The lifetime for these oxide CMC components is several times longer than for metals, which often deform. A further example is an oxide CMC lifting gate for a sintering furnace, which has survived more than 260,000 opening cycles.

Application in Brake Disk

Carbon/carbon (C/C) materials have found their way into the disk brakes of racing cars and aeroplanes, and C/SiC brake disks manufactured by the LSI process were qualified and are commercially available for luxury vehicles. The advantages of these C/SiC disks are:

- Very little wear, resulting in lifetime use for a car with a normal driving load of 300,000 km, is forecast by manufacturers.

- No fading is experienced, even under high load.

- No surface humidity effect on the friction coefficient shows up, as in C/C brake disks.

- The corrosion resistance, for example to the road salt, is much better than for metal disks.

- The disk mass is only 40% of a metal disk. This translates into less unsprung and rotating mass.

The weight reduction improves shock absorber response, road-holding comfort, agility, fuel economy, and thus driving comfort.

The SiC-matrix of LSI has a very low porosity, which protects the carbon fibres quite well. Brake disks do not experience temperatures above 500 °C for more than a few hours in their lifetime. Oxidation is therefore not a problem in this application. The reduction of manufacturing costs will decide the success of this application for middle-class cars.

Application in Slide Bearings

Conventional SiC, or sometimes the less expensive SiSiC, have been used successfully for more than 25 years in slide or journal bearings of pumps. The pumped liquid itself provides the lubricant for the bearing. Very good corrosion resistance against practically all kinds of media, and very low wear and low friction coefficients are the basis of this success. These bearings consist of a static bearing, shrink-fitted in its metallic environment, and a rotating shaft sleeve, mounted on the shaft. Under compressive stress the ceramic static bearing has a low risk of failure, but a SiC shaft sleeve does not have this situation and must therefore have a large wall thickness and/or be specially designed. In large pumps with shafts 100–350 mm in diameter, the risk of failure is higher due to the changing requirements on the pump performance – for example, load changes during operation. The introduction of SiC/SiC as a shaft sleeve material has proven to be very successful. Test rig experiments showed an almost triple specific load capability of the bearing system with a shaft sleeve made of SiC/SiC, sintered SiC as static bearing, and water at 80 °C as lubricant. The specific load capacity of a bearing is usually given in W/mm² and calculated as a product of the load (MPa), surface speed of the bearing (m/s) and friction coefficient; it is equal to the power loss of the bearing system due to friction.

Components for a ceramic slide bearing; the picture shows a sintered SiC-bearing for a hydrostatic slide bearing and a CVI-SiC/SiC-shaft sleeve shrink-fitted on metal, a system tested with liquid oxygen as lubricant.

In boiler feedwater pumps of power stations, which pump several thousand cubic meters of hot water to a level of 2000 m, and in tubular casing pumps for water works or sea water desalination plants (pumping up to 40,000 m³ to a level of around 20 m) this slide bearing concept, namely SiC/SiC shaft sleeve and SiC bearing, has been used since 1994.

This bearing system has been tested in pumps for liquid oxygen, for example in oxygen turbopumps for thrust engines of space rockets, with the following results. SiC and SiC/SiC are compatible with liquid oxygen. In an auto-ignition test according to the French standard NF 28-763, no auto-ignition was observed with powdered SiC/SiC in 20 bar pure oxygen at temperatures up to 525 °C. Tests have shown that the friction coefficient is half, and wear one fiftieth of standard metals used in this environment. A hydrostatic bearing system has survived several hours at a speed up to 10,000 revolutions per minute, various loads, and 50 cycles of start/stop transients without any significant traces of wear.

Other Applications and Developments

- Thrust control flaps for military jet engines

- Components for fusion and fission reactors

- Friction systems for various applications

- Nuclear applications

Transparent Ceramics

Many ceramic materials, both glassy and crystalline, have found use as optically transparent materials in various forms from bulk solid-state components to high surface area forms such as thin films, coatings, and fibers. Such devices have found widespread use for various applications in the electro-optical field including: optical fibers for guided lightwave transmission, optical switches, laser amplifiers and lenses, hosts for solid-state lasers and optical window materials for gas lasers, and infrared (IR) heat seeking devices for missile guidance systems and IR night vision.

Transparent spinel ($MgAl_2O_4$) ceramic is used traditionally for applications such as high-energy laser windows because of its excellent transmission in visible wavelengths and mid-wavelength infrared (0.2-5.0µm) when combined with selected materials - source: U.S. Naval Research Laboratory

While single-crystalline ceramics may be largely defect-free (particularly within the spatial scale of the incident light wave), optical transparency in polycrystalline materials is limited by the amount of light that is scattered by their microstructural features. The amount of light scattering therefore depends on the wavelength of the incident radiation, or light.

For example, since visible light has a wavelength scale on the order of hundreds of nanometers, scattering centers will have dimensions on a similar spatial scale. Most ceramic materials, such as alumina and its compounds, are formed from fine powders, yielding a fine grained polycrystalline microstructure that is filled with scattering centers comparable to the wavelength of visible light. Thus, they are generally opaque as opposed to transparent materials. Recent nanoscale technology, however, has made possible the production of (poly)crystalline transparent ceramics such as alumina Al_2O_3, yttria alumina garnet (YAG), and neodymium-doped Nd:YAG.

Introduction

Transparent ceramics have recently acquired a high degree of interest and notoriety. Basic applications include lasers and cutting tools, transparent armor windows, night vision devices (NVD), and nose cones for heat seeking missiles. Currently available infrared (IR) transparent materials typically exhibit a trade-off between optical performance and mechanical strength. For example, sapphire (crystalline alumina) is very strong, but lacks full transparency throughout the 3–5 micrometer mid-IR range. Yttria is fully transparent from 3–5 micrometers, but lacks sufficient strength, hardness, and thermal shock resistance for high-performance aerospace applications. Not surprisingly, a combination of these two materials in the form of the yttria-alumina garnet (YAG) has proven to be one of the top performers in the field.

Synthetic sapphire - single-crystal aluminum oxide (sapphire – Al_2O_3) is a transparent ceramic

In 1961, General Electric began selling transparent alumina Lucalox bulbs. In 1966, GE announced a ceramic "transparent as glass," called Yttralox. In 2004, Anatoly Rosenflanz and colleagues at 3M used a "flame-spray" technique to alloy aluminium oxide (or alumina) with rare-earth metal oxides in order to produce high strength glass-ceramics with good optical properties. The method avoids many of the problems encountered in conventional glass forming and may be extensible to other oxides. This goal has been readily accomplished and amply demonstrated in laboratories and research facilities worldwide using the emerging chemical processing methods encompassed by the methods of sol-gel chemistry and nanotechnology.

Many ceramic materials, both glassy and crystalline, have found use as hosts for solid-state lasers and as optical window materials for gas lasers. The first working laser was made by Theodore H. Maiman in 1960 at Hughes Research Laboratories in Malibu, who had the edge on other research teams led by Charles H. Townes at Columbia University, Arthur Schawlow at Bell Labs, and Gould at TRG (Technical Research Group). Maiman used a solid-state light-pumped synthetic ruby to produce red laser light at a wavelength of 694 nanometers (nm). Synthethic ruby lasers are still in use. Both sapphires and rubies are corundum, a crystalline form of aluminium oxide (Al2O3).

Crystals

Ruby lasers consist of single-crystal sapphire alumina (Al_2O_3) rods doped with a small concentration of chromium Cr, typically in the range of 0.05%. The end faces are highly polished with a planar and parallel configuration. Neodymium-doped YAG (Nd:YAG) has proven to be one of the best solid-state laser materials. Its indisputable dominance in a broad variety of laser applications is determined by a combination of high emission cross section with long spontaneous emission lifetime, high damage threshold, mechanical strength, thermal conductivity, and low thermal beam distortion. The fact that the Czochralski crystal growth of Nd:YAG is a matured, highly reproducible and relatively simple technological procedure adds significantly to the value of the material.

Nd:YAG lasers are used in manufacturing for engraving, etching, or marking a variety of metals and plastics. They are extensively used in manufacturing for cutting and welding steel and various alloys. For automotive applications (cutting and welding steel) the power levels are typically 1–5 kW. In addition, Nd:YAG lasers are used in ophthalmology to correct posterior capsular opacification, a condition that may occur after cataract surgery, and for peripheral iridotomy in patients with acute angle-closure glaucoma, where it has superseded surgical iridectomy. Frequency-doubled Nd:YAG lasers (wavelength 532 nm) are used for pan-retinal photocoagulation in patients with diabetic retinopathy. In oncology, Nd:YAG lasers can be used to remove skin cancers. These lasers are also used extensively in the field of cosmetic medicine for laser hair removal and the treatment of minor vascular defects such as spider veins on the face and legs. Recently used for dissecting cellulitis, a rare skin disease usually occurring on the scalp. Using hysteroscopy in the field of gynecology, the Nd:YAG laser has been used for removal of uterine septa within the inside of the uterus. In dentistry, Nd:YAG lasers are used for soft tissue surgeries in the oral cavity.

Currently, high powered Nd:glass lasers as large as a football field are used for inertial confinement fusion, nuclear weapons research, and other high energy density physics experiments

Glasses

Glasses (non-crystalline ceramics) also are used widely as host materials for lasers. Relative to crystalline lasers, they offer improved flexibility in size and shape and may be readily manufactured as large, homogeneous, isotropic solids with excellent optical properties. The indices of refraction of glass laser hosts may be varied between approximately 1.5 and 2.0, and both the temperature coefficient of n and the strain-optical coefficient may be tailored by altering the chemical composition. Glasses have lower thermal conductivities than the alumina or YAG, however, which imposes limitations on their use in continuous and high repetition-rate applications.

The principal differences between the behavior of glass and crystalline ceramic laser host materials are associated with the greater variation in the local environment of lasing ions in amorphous solids. This leads to a broadening of the fluorescent levels in glasses. For example, the width of the Nd^{3+} emission in YAG is ~ 10 angstroms as compared to ~ 300 angstroms in typical oxide glasses. The broadened fluorescent lines in glasses make it more difficult to obtain continuous wave laser operation (CW), relative to the same lasing ions in crystalline solid laser hosts.

Several glasses are used in transparent armor, such as normal plate glass (soda-lime-silica), borosilicate glass, and fused silica. Plate glass has been the most common glass used due to its low cost. But greater requirements for the optical properties and ballistic performance have necessitated the development of new materials. Chemical or thermal treatments can increase the strength of glasses, and the controlled crystallization of certain glass compositions can produce optical quality glass-ceramics. Alstom Grid Ltd. currently produces a lithium di-silicate based glass-ceramic known as TransArm, for use in transparent armor systems. It has all the workability of an amorphous glass, but upon recrystallization it demonstrates properties similar to a crystalline ceramic. Vycor is 96% fused silica glass, which is crystal clear, lightweight and high strength. One advantage of these type of materials is that they can be produced in large sheets and other curved shapes.

Nanomaterials

It has been shown fairly recently that laser elements (amplifiers, switches, ion hosts, etc.) made from fine-grained ceramic nanomaterials—produced by the low temperature sintering of high purity nanoparticles and powders—can be produced at a relatively low cost. These components are free of internal stress or intrinsic birefringence, and allow relatively large doping levels or optimized custom-designed doping profiles. This highlights the use of ceramic nanomaterials as being particularly important for high-energy laser elements and applications.

Primary scattering centers in polycrystalline nanomaterials—made from the sintering of high purity nanoparticles and powders—include microstructural defects such as residual porosity and grain boundaries. Thus, opacity partly results from the incoherent scattering of light at internal surfaces and interfaces. In addition to porosity, most of the interfaces or internal surfaces in ceramic nanomaterials are in the form of grain boundaries which separate nanoscale regions of crystalline order. Moreover, when the size of the scattering center (or grain boundary) is reduced well below the size of the wavelength of the light being scattered, the light scattering no longer occurs to any significant extent.

In the processing of high performance ceramic nanomaterials with superior opto-mechanical properties under adverse conditions, the size of the crystalline grains is determined largely by the size of the crystalline particles present in the raw material during the synthesis or formation of the object. Thus a reduction of the original particle size well below the wavelength of visible light (~ 0.5 μm or 500 nm) eliminates much of the light scattering, resulting in a translucent or even transparent material.

Furthermore, results indicate that microscopic pores in sintered ceramic nanomaterials, mainly trapped at the junctions of microcrystalline grains, cause light to scatter and prevented true transparency. It has been observed that the total volume fraction of these nanoscale pores (both intergranular and intragranular porosity) must be less than 1% for high-quality optical transmission, i.e. the density has to be 99.99% of the theoretical crystalline density.

Lasers

Nd:YAG

For example, a 1.46 kW Nd:YAG laser has been demonstrated by Konoshima Chemical Co. in Japan. In addition, Livermore researchers realized that these fine-grained ceramic nanomaterials might greatly benefit high-powered lasers used in the National Ignition Facility (NIF) Programs Directorate. In particular, a Livermore research team began to acquire advanced transparent nanomaterials from Konoshima to determine if they could meet the optical requirements needed for Livermore's Solid-State Heat Capacity Laser (SSHCL). Livermore researchers have also been testing applications of these materials for applications such as advanced drivers for laser-driven fusion power plants.

Assisted by several workers from the NIF, the Livermore team has produced 15 mm diameter samples of transparent Nd:YAG from nanoscale particles and powders, and determined the most important parameters affecting their quality. In these objects, the team largely followed the Japanese production and processing methodologies, and used an in house furnace to vacuum sinter the nanopowders. All specimens were then sent out for hot isostatic pressing (HIP). Finally, the components were returned to Livermore for coating and testing, with results indicating exceptional optical quality and properties.

One Japanese/East Indian consortium has focused specifically on the spectroscopic and stimulated emission characteristics of Nd^{3+} in transparent YAG nanomaterials for laser applications. Their materials were synthesized using vacuum sintering techniques. The spectroscopic studies suggest overall improvement in absorption and emission and reduction in scattering loss. Scanning electron microscope and transmission electron microscope observations revealed an excellent optical quality with low pore volume and narrow grain boundary width. Fluorescence and Raman measurements reveal that the Nd^{3+} doped YAG nanomaterial is comparable in quality to its single-crystal counterpart in both its radiative and non-radiative properties. Individual Stark levels are obtained from the absorption and fluorescence spectra and are analyzed in oredr to identify the stimulated emission channels possible in the material. Laser performance studies favor the use of high dopant concentration in the design of an efficient microchip laser. With 4 at% dopant, the group obtained a slope efficiency of 40%. High-power laser experiments yield an optical-to-optical conversion efficiency of 30% for Nd (0.6 at%) YAG nanomaterial as compared to 34% for an Nd (0.6 at%) YAG single crystal. Optical gain measurements conducted in these materials also show values comparable to single crystal, supporting the contention that these materials could be suitable substitutes to single crystals in solid-state laser applications.

Yttria, Y_2O_3

The initial work in developing transparent yttrium oxide nanomaterials was carried out by General Electric in the 1960s.

In 1966, a transparent ceramic, Yttralox, was invented by Dr. Richard C. Anderson at the General Electric Research Laboratory, with further work at GE's Metallurgy and Ceramics Laboratory by Drs. Paul J. Jorgensen, Joseph H. Rosolowski, and Douglas St. Pierre. Yttralox is "transparent as glass," has a melting point twice as high, and transmits frequencies in the near infrared band as well as visible light.

IR 100 Award, Yttralox, 1967

Gemstones of Yttralox transparent ceramic

Richard C. Anderson holding a sample of Yttralox

Further development of yttrium ceramic nanomaterials was carried out by General Electric in the 1970s in Schenectady and Cleveland, motivated by lighting and ceramic laser applications. Yttralox, transparent yttrium oxide Y_2O_3 containing ~ 10% thorium oxide (ThO_2) was fabricated by Greskovich and Woods. The additive served to control grain growth during densification, so that porosity remained on grain boundaries and not trapped inside grains where it would be quite difficult to eliminate during the initial stages of sintering. Typically, as polycrystalline ceramics densify during heat treatment, grains grow in size while the remaining porosity decreases both in volume fraction and in size. Optically transparent ceramics must be virtually pore-free.

GE's transparent Yttralox was followed by GTE's lanthana-doped yttria with similar level of additive. Both of these materials required extended firing times at temperatures above 2000 °C. La_2O_3 – doped Y_2O_3 is of interest for infrared (IR) applications because it is one of the longest wavelength transmitting oxides. It is refractory with a melting point of 2430 °C and has a moderate coefficient of thermal expansion coefficient. The thermal shock and erosion resistance is considered to be intermediate among the oxides, but outstanding compared to non-oxide IR transmitting materials. A major consideration is the low emissivity of yttria, which limits background radiation upon heating. It is also known that the phonon edge gradually moves to shorter wavelengths as a material is heated.

In addition, ytrria itself, Y_2O_3 has been clearly identified as a prospective solid-state laser material. In particular, lasers with ytterbium as dopant allow the efficient operation both in cw operation and in pulsed regimes.

At high concentration of excitations (of order of 1%) and poor cooling, the quenching of emission at laser frequency and avalanche broadband emission takes place.

Future

The Livermore team is also exploring new ways to chemically synthesize the initial nanopowders. Borrowing on expertise developed in CMS over the past 5 years, the team is synthesizing nano-powders based on sol-gel processing, and then sintering them accordingly in order to obtain the solid-state laser components. Another technique being tested utilizes a combustion process in order to generate the powders by burning an organic solid containing yttrium, aluminum, and neo-dymium. The smoke is then collected, which consists of spherical nanoparticles.

The Livermore team is also exploring new forming techniques (e.g. extrusion molding) which have the capacity to create more diverse, and possibly more complicated, shapes. These include shells and tubes for improved coupling to the pump light and for more efficient heat transfer. In addition, different materials can be co-extruded and then sintered into a monolithic transparent solid. An amplifier slab can formed so that part of the structure acts in guided lightwave transmission in order to focus pump light from laser diodes into regions with a high concentration of dopant ions near the slab center.

In general, nanomaterials promise to greatly expand the availability of low-cost, high-end laser components in much larger sizes than would be possible with traditional single crystalline ceram-ics. Many classes of laser designs could benefit from nanomaterial-based laser structures such as amplifies with built-in edge claddings. Nanomaterials could also provide more robust and compact designs for high-peak power, fusion-class lasers for stockpile stewardship, as well as high-aver-age-power lasers for global theater ICBM missile defense systems (e.g. Strategic Defense Initiative SDI, or more recently the Missile Defense Agency.

Night Vision

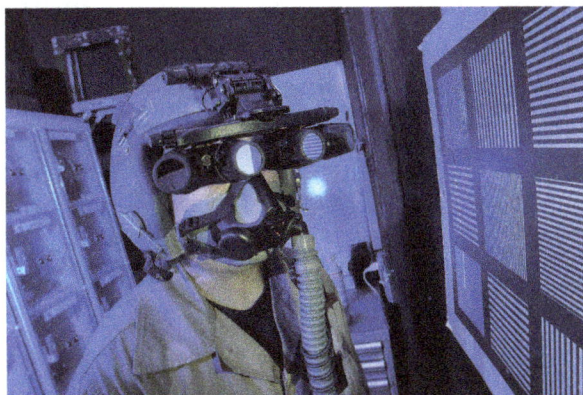

Panoramic Night Vision Goggles in testing.

A night vision device (NVD) is an optical instrument that allows images to be produced in levels of light approaching total darkness. They are most often used by the military and law enforcement agencies, but are available to civilian users. Night vision devices were first used in World War II, and came into wide use during the Vietnam War. The technology has evolved greatly since their introduction, leading to several "generations" of night vision equipment with performance increasing and price decreasing. The United States Air Force is experimenting with Panoramic Night Vision Goggles (PNVGs) which double the user's field of view to approximately 95 degrees by using four 16 mm image intensifiers tubes, rather than the more standard two 18 mm tubes.

Thermal images are visual displays of the amount of infrared (IR) energy emitted, transmitted, and reflected by an object. Because there are multiple sources of the infrared energy, it is difficult to get an accurate temperature of an object using this method. A thermal imaging camera is capable of performing algorithms to interpret that data and build an image. Although the image shows the viewer an approximation of the temperature at which the object is operating, the camera is using multiple sources of data based on the areas surrounding the object to determine that value rather than detecting the temperature.

Night vision infrared devices image in the near-infrared, just beyond the visual spectrum, and can see emitted or reflected near-infrared in complete visual darkness. All objects above the absolute zero temperature (0 K) emit infrared radiation. Hence, an excellent way to measure thermal variations is to use an infrared vision device, usually a focal plane array (FPA) infrared camera capable of detecting radiation in the mid (3 to 5 μm) and long (7 to 14 μm) wave infrared bands, denoted as MWIR and LWIR, corresponding to two of the high transmittance infrared windows. Abnormal temperature profiles at the surface of an object are an indication of a potential problem. Infrared thermography, thermal imaging, and thermal video, are examples of infrared imaging science. Thermal imaging cameras detect radiation in the infrared range of the electromagnetic spectrum (roughly 900–14,000 nanometers or 0.9–14 μm) and produce images of that radiation, called *thermograms*.

Since infrared radiation is emitted by all objects near room temperature, according to the black body radiation law, thermography makes it possible to see one's environment with or without visible illumination. The amount of radiation emitted by an object increases with temperature. Therefore, thermography allows one to see variations in temperature. When viewed through a thermal imaging camera, warm objects stand out well against cooler backgrounds; humans and other warm-blooded animals become easily visible against the environment, day or night. As a result, thermography is particularly useful to the military and to security services.

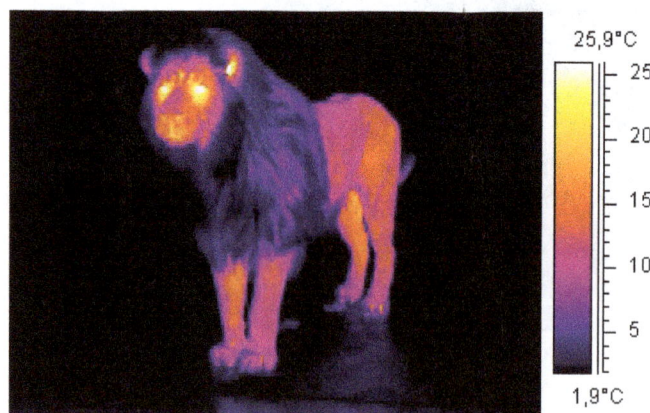

Thermogram of a lion

Thermography

In thermographic imaging, infrared radiation with wavelengths between 8–13 micrometers strikes the detector material, heating it, and thus changing its electrical resistance. This resistance change is measured and processed into temperatures which can be used to create an image. Unlike other types of infrared detecting equipment, microbolometers utilizing a transparent ceramic detector do not require cooling. Thus, a microbolometer is essentially an uncooled thermal sensor.

The material used in the detector must demonstrate large changes in resistance as a result of minute changes in temperature. As the material is heated, due to the incoming infrared radiation, the resistance of the material decreases. This is related to the material's temperature coefficient of resistance (TCR) specifically its negative temperature coefficient. Industry currently manufactures microbolometers that contain materials with TCRs near -2%.

VO_2 and V_2O_5

The most commonly used ceramic material in IR radiation microbolometers is vanadium oxide. The various crystalline forms of vanadium oxide include both VO_2 and V_2O_5. Deposition at high temperatures and performing post-annealing allows for the production of thin films of these crystalline compounds with superior properties, which may be easily integrated into the fabrication process. VO_2 has low resistance but undergoes a metal-insulator phase change near 67 °C and also has a lower TCR value. On the other hand, V_2O_5 exhibits high resistance and also high TCR.

Other IR transparent ceramic materials that have been investigated include doped forms of CuO, MnO and SiO.

Missiles

AIM-9 Sidewinder	
Place of origin	United States

Many ceramic nanomaterials of interest for transparent armor solutions are also used for electromagnetic (EM) windows. These applications include radomes, IR domes, sensor protection, and multi-spectral windows. Optical properties of the materials used for these applications are critical, as the transmission window and related cut-offs (UV – IR) control the spectral bandwidth over which the window is operational. Not only must these materials possess abrasion resistance and strength properties common of most armor applications, but due to the extreme temperatures associated with the environment of military aircraft and missiles, they must also possess excellent thermal stability.

Thermal radiation is electromagnetic radiation emitted from the surface of an object which is due to the object's temperature. Infrared homing refers to a passive missile guidance system which uses the emission from a target of electromagnetic radiation in the infrared part of the spectrum to track it. Missiles that use infrared seeking are often referred to as "heat-seekers", since infrared is just below the visible spectrum of light in frequency and is radiated strongly by hot bodies. Many objects such as people, vehicle engines and aircraft generate and retain heat, and as such, are especially visible in the infrared wavelengths of light compared to objects in the background.

Sapphire

The current material of choice for high-speed infrared-guided missile domes is single-crystal sapphire. The optical transmission of sapphire does not extend to cover the entire mid-infrared range (3–5 μm), but starts to drop off at wavelengths greater than approximately 4.5 μm at room temperature. While the strength of sapphire is better than that of other available mid-range infrared dome materials at room temperature, it weakens above ~600 °C.

Limitations to larger area sapphires are often business related, in that larger induction furnaces and costly tooling dies are necessary in order to exceed current fabrication limits. However, as an industry, sapphire producers have remained competitive in the face of coating-hardened glass and new ceramic nanomaterials, and still managed to offer high performance and an expanded market.

Yttria, Y_2O_3

Alternative materials, such as yttrium oxide, offer better optical performance, but inferior mechanical durability. Future high-speed infrared-guided missiles will require new domes that are substantially more durable than those in use today, while still retaining maximum transparency across a wide wavelength range. A long-standing trade-off exists between optical bandpass and mechanical durability within the current collection of single-phase infrared transmitting materials, forcing missile designers to compromise on system performance. Optical nanocomposites may present the opportunity to engineer new materials that overcome this traditional compromise.

The first full scale missile domes of transparent yttria manufactured from nanoscale ceramic powders were developed in the 1980s under Navy funding. Raytheon perfected and characterized its undoped polycrystalline yttria, while lanthana-doped yttria was similarly developed by GTE Labs. The two versions had comparable IR transmittance, fracture toughness, and thermal expansion, while the undoped version exhibited twice the value of thermal conductivity.

Renewed interest in yttria windows and domes has prompted efforts to enhance mechanical properties by using nanoscale materials with submicrometer or nanosized grains. In one study, three vendors were selected to provide nanoscale powders for testing and evaluation, and they were compared to a conventional (5 μm) yttria powder previously used to prepare transparent yttria. While all of the nanopowders evaluated had impurity levels that were too high to allow processing to full transparency, 2 of them were processed to theoretical density and moderate transparency. Samples were sintered to a closed pore state at temperatures as low as 1400 C.

After the relatively short sintering period, the component is placed in a hot isostatic press (HIP) and processed for 3 – 10 hours at ~ 30 kpsi(~200 MPa) at a temperature similar to that of the initial sintering. The applied isostatic pressure provides additional driving force for densification by substantially increasing the atomic diffusion coefficients, which promotes additional viscous flow at or near grain boundaries and intergranular pores. Using this method, transparent yttria nanomaterials were produced at lower temperatures, shorter total firing times, and without extra additives which tend to reduce the thermal conductivity.

Recently, a newer method has been devleoped by Mouzon, which relies on the methods of glass-encapsulation, combined with vacuum sintering at 1600 °C followed by hot isostatic pressing (HIP) at 1500 °C of a highly agglomerated commercial powder. The use of evacuated glass capsules to perform HIP treatment allowed samples that showed open porosity after vacuum sintering to be sintered to transparency. The sintering response of the investigated powder was studied by careful microstructural observations using scanning electron microscopy and optical microscopy both in reflection and transmission. The key to this method is to keep porosity intergranular during pre-sintering, so that it can be removed subsequently by HIP treatment. It was found that agglomerates of closely packed particles are helpful to reach that purpose, since they densify fully and leave only intergranular porosity.

Composites

Prior to the work done at Raytheon, optical properties in nanocomposite ceramic materials had received little attention. Their studies clearly demonstrated near theoretical transmission in nanocomposite optical ceramics for the first time. The yttria/magnesia binary system is an ideal model system for nanocomposite formation. There is limited solid solubility in either one of the constituent phases, permitting a wide range of compositions to be investigated and compared to each other. According to the phase diagram, bi-phase mixtures are stable for all temperatures below ~ 2100 °C. In addition, neither yttria nor magnesia shows any absorption in the 3 – 5 μm mid-range IR portion of the EM spectrum.

In optical nanocomposites, two or more interpenetrating phases are mixed in a sub-micrometer grain sized, fully dense body. Infrared light scattering can be minimized (or even eliminated) in the material as long as the grain size of the individual phases is significantly smaller than infrared wavelengths. Experimental data suggests that limiting the grain size of the nanocomposite to approximately 1/15th of the wavelength of light is sufficient to limit scattering.

Nanocomposites of yttria and magnesia have been produced with a grain size of approximately 200 nm. These materials have yielded good transmission in the 3–5 μm range and strengths higher than that for single-phase individual constituents. Enhancement of mechanical properties in nanocomposite ceramic materials has been extensively studied. Significant increases in strength (2–5 times), toughness (1–4 times), and creep resistance have been observed in systems including SiC/Al_2O_3, SiC/Si_3N_4, SiC/MgO, and Al_2O_3/ZrO_2.

The strengthening mechanisms observed vary depending on the material system, and there does not appear to be any general consensus regarding strengthening mechanisms, even within a given system. In the SiC/Al_2O_3 system, for example, it is widely known and accepted that the addition of SiC particles to the Al_2O_3 matrix results in a change of failure mechanism from intergranular (between grains) to intragranular (within grains) fracture. The explanations for improved strength include:

- A simple reduction of processing flaw concentration during nanocomposite fabrication.

- Reduction of the critical flaw size in the material—resulting in increased strength as predicted by the Hall-Petch relation)

- Crack deflection at nanophase particels due to residual thermal stresses introduced upon cooling form processing temperatures.

- Microcracking along stress-induced dislocations in the matrix material.

Armor

There is an increasing need in the military sector for high-strength, robust materials which have the capability to transmit light around the visible (0.4–0.7 micrometers) and mid-infrared (1–5 micrometers) regions of the spectrum. These materials are needed for applications requiring transparent armor. Transparent armor is a material or system of materials designed to be optically transparent, yet protect from fragmentation or ballistic impacts. The primary requirement for a transparent armor system is to not only defeat the designated threat but also provide a multi-hit capability with minimized distortion of surrounding areas. Transparent armor windows must also be compatible with night vision equipment. New materials that are thinner, lightweight, and offer better ballistic performance are being sought.

Existing transparent armor systems typically have many layers, separated by polymer (e.g. poly-carbonate) interlayers. The polymer interlayer is used to mitigate the stresses from thermal expansion mismatches, as well as to stop crack propagation from ceramic to polymer. The polycarbonate is also currently used in applications such as visors, face shields and laser protection goggles. The search for lighter materials has also led to investigations into other polymeric materials such as transparent nylons, polyurethane, and acrylics. The optical properties and durability of transparent plastics limit their use in armor applications. Investigations carried out in the 1970s had shown promise for the use of polyurethane as armor material, but the optical properties were not adequate for transparent armor applications.

Several glasses are utilized in transparent armor, such as normal plate glass (soda-lime-silica), borosilicate glasses, and fused silica. Plate glass has been the most common glass used due to its low cost, but greater requirements for the optical properties and ballistic performance have generated the need for new materials. Chemical or thermal treatments can increase the strength of glasses, and the controlled crystallization of certain glass systems can produce transparent glass-ceramics. Alstom Grid Research & Technology (Stafford, UK), produced a lithium disilicate based glass-ceramic known as TransArm, for use in transparent armor systems with continuous production yielding vehicle windscreen sized pieces (and larger). The inherent advantages of glasses and glass-ceramics include having lower cost than most other ceramic materials, the ability to be produced in curved shapes, and the ability to be formed into large sheets.

Transparent crystalline ceramics are used to defeat advanced threats. Three major transparent candidates currently exist: aluminum oxynitride (AlON), magnesium aluminate spinel (spinel), and single crystal aluminum oxide (sapphire). Aluminum oxynitride spinel ($Al_{23}O_{27}N_5$), one of the leading candidates for transparent armor, is produced by Surmet Corporation as AlON and marketed under the trade name ALON. The incorporation of nitrogen into an aluminum oxide stabilizes a spinel phase, which due to its cubic crystal structure, is an isotropic material that can be produced as a transparent polycrystalline material. Polycrystalline materials can be produced in complex geometries using conventional ceramic forming techniques such as pressing, (hot) isostatic pressing, and slip casting.

Aluminium Oxynitride Spinel

Aluminium oxynitride spinel ($Al_{23}O_{27}N_5$), abbreviated as AlON, is one of the leading candidates for transparent armor. It is produced by the Surmet Corporation under the trademark ALON. The incorporation of nitrogen into aluminium oxide stabilizes a crystalline spinel phase, which due to its cubic crystal structure and unit cell, is an isotropic material which can be produced as transparent ceramic nanomaterial. Thus, fine-grained polycrystalline nanomaterials can be produced and formed into complex geometries using conventional ceramic forming techniques such as hot pressing and slip casting.

The Surmet Corporation has acquired Raytheon's ALON business and is currently building a market for this technology in the area of Transparent Armor, Sensor windows, Reconnaissance windows and IR Optics such as Lenses and Domes and as an alternative to quartz and sapphire in the semiconductor market. The AlON based transparent armor has been tested to stop multi-hit threats including of 30calAPM2 rounds and 50calAPM2 rounds successfully. The high hardness of AlON provides a scratch resistance which exceeds even the most durable coatings for glass scanner windows, such as those used in supermarkets. Surmet has successfully produced a 15"x18" curved AlON window and is currently attempting to scale up the technology and reduce the cost. In addition, the U.S. Army and U.S. Air Force are both seeking development into next generation applications.

Spinel

Magnesium aluminate spinel ($MgAl_2O_4$) is a transparent ceramic with a cubic crystal structure with an excellent optical transmission from 0.2 to 5.5 micrometers in its polycrystalline form. Optical quality transparent spinel has been produced by sinter/HIP, hot pressing, and hot press/HIP operations, and it has been shown that the use of a hot isostatic press can improve its optical and physical properties.

Spinel offers some processing advantages over AlON, such as the fact that spinel powder is available from commercial manufacturers while AlON powders are proprietary to Raytheon. It is also capable of being processed at much lower temperatures than AlON and has been shown to possess superior optical properties within the infrared (IR) region. The improved optical characteristics make spinel attractive in sensor applications where effective communication is impacted by the protective missile dome's absorption characteristics.

Spinel shows promise for many applications, but is currently not available in bulk form from any manufacturer, although efforts to commercialize spinel are underway. The spinel products business is being pursued by two key U.S. manufacturers: "Technology Assessment and Transfer" and the "Surmet Corporation".

An extensive NRL review of the literature has indicated clearly that attempts to make high-quality spinel have failed to date because the densification dynamics of spinel are poorly understood. They have conducted extensive research into the dynamics involved during the densification of spinel. Their research has shown that LiF, although necessary, also has extremely adverse effects during the final stages of densification. Additionally, its distribution in the precursor spinel powders is of critical importance.

Traditional bulk mixing processes used to mix LiF sintering aid into a powder leave fairly inhomogeneous distribution of Lif that must be homogenized by extended heat treatment times at elevated temperatures. The homogenizing temperature for Lif/Spinel occurs at the temperature of fast reaction between the LiF and the Al_2O_3. In order to avoid this detrimental reaction, they have developed a new process that uniformly coats the spinel particles with the sintering aid. This allows them to reduce the amount of Lif necessary for densification and to rapidly heat through the temperature of maximum reactivity. These developments have allowed NRL to fabricate $MgAl_2O_4$ spinel to high transparency with extremely high reproducibility that should enable military as well as commercial use of spinel.

Sapphire

Single-crystal aluminum oxide (sapphire – Al_2O_3) is a transparent ceramic. Sapphire's crystal structure is rhombohedral and thus its properties are anisotropic, varying with crystallographic orientation. Transparent alumina is currently one of the most mature transparent ceramics from a production and application perspective, and is available from several manufacturers. But the cost is high due to the processing temperature involved, as well as machining costs to cut parts out of single crystal boules. It also has a very high mechanical strength – but that is dependent on the surface finish.

The high level of maturity of sapphire from a production and application standpoint can be attributed to two areas of business: electromagnetic spectrum windows for missiles and domes, and electronic/semiconductor industries and applications.

There are current programs to scale-up sapphire grown by the heat exchanger method or edge defined film-fed growth (EFG) processes. Its maturity stems from its use as windows and in semiconductor industry. Crystal Systems Inc. which uses single crystal growth techniques, is currently scaling their sapphire boules to 13-inch (330 mm) diameter and larger. Another producer, the Saint-Gobain Group produces transparent sapphire using an edge, defined growth technique. Sapphire grown by this technique produces an optically inferior material to that which is grown via single crystal techniques, but is much less expensive, and retains much of the hardness, transmission, and scratch-resistant characteristics. Saint-Gobain is currently capable of producing 0.43" thick (as grown) sapphire, in 12" × 18.5" sheets, as well as thick, single-curved sheets. The U.S. Army Research Laboratory is currently investigating use of this material in a laminate design for transparent armor systems. The Saint Gobain Group have commercialized the capability to meet flight requirements on the F-35 Joint Strike Fighter and F-22 Raptor next generation fighter aircraft.

Composites

Future high-speed infrared-guided missiles will require new dome materials that are substantially more durable than those in use today, while retaining maximum transparency across the entire operational spectrum or bandwidth. A long-standing compromise exists between optical bandpass and mechanical durability within the current group of single-phase (crystalline or glassy) IR transmitting ceramic materials, forcing missile designers to accept substandard overall system performance. Optical nanocomposites may provide the opportunity to engineer new materials that may overcome these traditional limitations.

For example, transparent ceramic armor consisting of a lightweight composite has been formed by utilizing a face plate of transparent alumina Al_2O_3 (or magnesia MgO) with a back-up plate of transparent plastic. The two plates (bonded together with a transparent adhesive) afford complete ballistic protection against 0.30 AP M2 projectiles at 0° obliquity with a muzzle velocity of 2,770 ft (840 m) per second. Another transparent composite armor provided complete protection for small arms projectiles up to and including caliber .50 AP M2 projectiles consisting of two or more layers of transparent ceramic material.

Nanocomposites of yttria and magnesia have been produced with an average grain size of ~200 nm. These materials have exhibited near theoretical transmission in the $3 - 5$ μm IR band. Additionally, such composites have yielded higher strengths than those observed for single phase solid-state components. Despite a lack of agreement regarding mechanism of failure, it is widely accepted that nanocomposite ceramic materials can and do offer improved mechanical properties over those of single phase materials or nanomaterials of uniform chemical composition.

It should also be noted here that nanocomposite ceramic materials also offer interesting mechanical properties not achievable in other materials, such as superplastic flow and metal-like machinability. It is anticipated that further development will result in high strength, high transparency nanomaterials which are suitable for application as next generation armor.

References

- Greenwood, N. N.; & Earnshaw, A. (1997). Chemistry of the Elements (2nd Edn.), Oxford:Butterworth-Heinemann. ISBN 0-7506-3365-4.

- Gray, Theodore (2012-04-03). The Elements: A Visual Exploration of Every Known Atom in the Universe. Black Dog & Leventhal Publishers. ISBN 9781579128951. Retrieved 6 May 2014.

- Greenwood, Norman N.; Earnshaw, Alan (1997). Chemistry of the Elements (2nd ed.). Butterworth-Heinemann. p. 149. ISBN 0-08-037941-9.

- Nishi, Yoshio; Doering, Robert (2000). Handbook of semiconductor manufacturing technology. CRC Press. pp. 324–325. ISBN 0-8247-8783-8.

- Peng, Hong (2004). Spark Plasma Sintering of Si3N4-based Ceramics: Sintering mechanism-Tailoring microstructure-Evaluating properties (PhD thesis). Stockholm University. ISBN 978-91-7265-834-9.[page needed]

- Antram, Nicholas; Morrice, Richard (2008). Brighton and Hove. Pevsner Architectural Guides. London: Yale University Press. ISBN 978-0-300-12661-7.

- N. Miriyala; J. Kimmel; J. Price; H. Eaton; G. Linsey; E. Sun (2002). "The evaluation of CFCC Liners After Field Testing in a Gas Turbine – III" (PDF): 109–118. doi:10.1115/GT2002-30585. ISBN 0-7918-3609-6.

- K.L. More; P.F. Tortorelli; L.R. Walker; J.B. Kimmel; N. Miriyala; J.R. Price; H.E. Eaton; E. Y. Sun; G.D. Linsey (2002). "Volume 4: Turbo Expo 2002, Parts A and B" (PDF): 155–162. doi:10.1115/GT2002-30630. ISBN 0-7918-3609-6.

- W. Krenkel, R. Renz, CMCs for Friction Applications, in Ceramic Matrix Composites, W. Krenkel editor, Wiley-VCH, 2008. ISBN 978-3-527-31361-7, p. 396

- P. Boullon; G. Habarou; P.C. Spriet; J.L. Lecordix; G.C. Ojard; G.D. Linsey; D.T. Feindel (2002). "Volume 4: Turbo Expo 2002, Parts A and B": 15–21. doi:10.1115/GT2002-30458. ISBN 0-7918-3609-6.

- N.P. Bansal, J.Lamon (ed.): "Ceramic Matrix Composites: Materials, Modeling and Technology". Wiley, Hoboken, NJ 2015. ISBN 978-1-118-23116-6, p. 609

- Advances in Ceramic Armor IV. Part I: Transparent Glasses and Ceramics], Ceramic Engineering and Science Proceedings, Vol. 29 (Wiley, American Ceramic Society, 2008) ISBN 0-470-34497-0

Strength of Ceramics: A Comprehensive Study

Grain-boundary strengthening is a technique of strengthening materials by basically changing their average grain size. The alternative methods used for strengthening ceramics are strengthening mechanisms of materials and solid solution strengthening. This chapter is an overview of the subject matter incorporating all the major aspects of strength of ceramics.

Grain Boundary Strengthening

Grain-boundary strengthening (or Hall–Petch strengthening) is a method of strengthening materials by changing their average crystallite (grain) size. It is based on the observation that grain boundaries impede dislocation movement and that the number of dislocations within a grain have an effect on how easily dislocations can traverse grain boundaries and travel from grain to grain. So, by changing grain size one can influence dislocation movement and yield strength. For example, heat treatment after plastic deformation and changing the rate of solidification are ways to alter grain size.

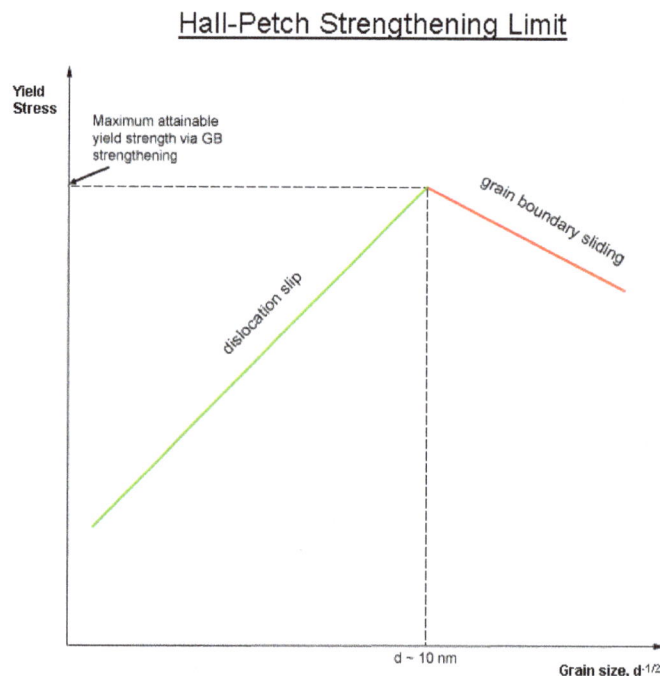

Hall–Petch Strengthening is limited by the size of dislocations. Once the grain size reaches about 10 nanometres (3.9×10^{-7} in), grain boundaries start to slide.

Theory

In grain-boundary strengthening, the grain boundaries act as pinning points impeding further dislocation propagation. Since the lattice structure of adjacent grains differs in orientation, it requires more energy for a dislocation to change directions and move into the adjacent grain. The grain boundary is also much more disordered than inside the grain, which also prevents the dislocations from moving in a continuous slip plane. Impeding this dislocation movement will hinder the onset of plasticity and hence increase the yield strength of the material.

Under an applied stress, existing dislocations and dislocations generated by Frank-Read Sources will move through a crystalline lattice until encountering a grain boundary, where the large atomic mismatch between different grains creates a repulsive stress field to oppose continued dislocation motion. As more dislocations propagate to this boundary, dislocation 'pile up' occurs as a cluster of dislocations are unable to move past the boundary. As dislocations generate repulsive stress fields, each successive dislocation will apply a repulsive force to the dislocation incident with the grain boundary. These repulsive forces act as a driving force to reduce the energetic barrier for diffusion across the boundary, such that additional pile up causes dislocation diffusion across the grain boundary, allowing further deformation in the material. Decreasing grain size decreases the amount of possible pile up at the boundary, increasing the amount of applied stress necessary to move a dislocation across a grain boundary. The higher the applied stress needed to move the dislocation, the higher the yield strength. Thus, there is then an inverse relationship between grain size and yield strength, as demonstrated by the Hall–Petch equation. However, when there is a large direction change in the orientation of the two adjacent grains, the dislocation may not necessarily move from one grain to the other but instead create a new source of dislocation in the adjacent grain. The theory remains the same that more grain boundaries create more opposition to dislocation movement and in turn strengthens the material.

Obviously, there is a limit to this mode of strengthening, as infinitely strong materials do not exist. Grain sizes can range from about 100 μm (0.0039 in) (large grains) to 1 μm (3.9×10⁻⁵ in) (small grains). Lower than this, the size of dislocations begins to approach the size of the grains. At a grain size of about 10 nm (3.9×10⁻⁷ in), only one or two dislocations can fit inside a grain. This scheme prohibits dislocation pile-up and instead results in grain boundary diffusion. The lattice resolves the applied stress by grain boundary sliding, resulting in a *decrease* in the material's yield strength.

To understand the mechanism of grain boundary strengthening one must understand the nature of dislocation-dislocation interactions. Dislocations create a stress field around them given by:

$$\sigma \propto \frac{Gb}{r},$$

where G is the material's shear modulus, b is the Burgers vector, and r is the distance from the dislocation. If the dislocations are in the right alignment with respect to each other, the local stress fields they create will repel each other. This helps dislocation movement along grains and across grain boundaries. Hence, the more dislocations are present in a grain, the greater the stress field felt by a dislocation near a grain boundary:

$$\tau_{felt} = \tau_{applied} + n_{dislocation}\tau_{dislocation}$$

Subgrain Strengthening

A subgrain is a part of the grain that is only slightly disoriented from other parts of the grain. Current research is being done to see the effect of subgrain strengthening in materials. Depending on the processing of the material, subgrains can form within the grains of the material. For example, when Fe-based material is ball-milled for long periods of time (e.g. 100+ hours), subgrains of 60–90 nm are formed. It has been shown that the higher the density of the subgrains, the higher the yield stress of the material due to the increased subgrain boundary. The strength of the metal was found to vary reciprocally with the size of the subgrain, which is analogous to the Hall–Petch equation. The subgrain boundary strengthening also has a breakdown point of around a subgrain size of 0.1 μm, which is the size where any subgrains smaller than that size would decrease yield strength. .

Hall–Petch Relationship

Hall–Petch constants		
Material	σ_o [MPa]	k [MPa m$^{1/2}$]
Copper	25	0.11
Titanium	80	0.40
Mild steel	70	0.74
Ni$_3$Al	300	1.70

There is an inverse relationship between delta yield strength and grain size to some power, x.

$$\Delta \tau \propto \frac{k}{d^x}$$

where k is the strengthening coefficient and both k and x are material specific. The smaller the grain size, the smaller the repulsion stress felt by a grain boundary dislocation and the higher the applied stress needed to propagate dislocations through the material.

The relation between yield stress and grain size is described mathematically by the Hall–Petch equation:

$$\sigma_y = \sigma_0 + \frac{k_y}{\sqrt{d}}$$

where σ_y is the yield stress, σ_o is a materials constant for the starting stress for dislocation movement (or the resistance of the lattice to dislocation motion), k_y is the strengthening coefficient (a constant specific to each material), and d is the average grain diameter.

Theoretically, a material could be made infinitely strong if the grains are made infinitely small. This is impossible though, because the lower limit of grain size is a single unit cell of the material. Even then, if the grains of a material are the size of a single unit cell, then the material is in fact amorphous, not crystalline, since there is no long range order, and dislocations can not be defined in an amorphous material. It has been observed experimentally that the microstructure with the highest yield strength is a grain size of about 10 nm (3.9×10^{-7} in), because grains smaller than this undergo another yielding mechanism, grain boundary sliding. Producing engineering materials with this ideal grain size is difficult because only thin films can be reliably produced with grains of this size.

History

In the early 1950s two groundbreaking series of papers were written independently on the relationship between grain boundaries and strength.

In 1951, while at the University of Sheffield, E. O. Hall wrote three papers which appeared in volume 64 of the Proceedings of the Physical Society. In his third paper, Hall showed that the length of slip bands or crack lengths correspond to grain sizes and thus a relationship could be established between the two. Hall concentrated on the yielding properties of mild steels.

Based on his experimental work carried out in 1946–1949, N. J. Petch of the University of Leeds, England published a paper in 1953 independent from Hall's. Petch's paper concentrated more on brittle fracture. By measuring the variation in cleavage strength with respect to ferritic grain size at very low temperatures, Petch found a relationship exact to that of Hall's. Thus this important relationship is named after both Hall and Petch.

Reverse or Inverse Hall–Petch Relation

The Hall–Petch relation predicts that as the grain size decreases the yield strength increases. The Hall–Petch relation was experimentally found to be an effective model for materials with grain sizes ranging from 1 millimeter to 1 micrometer. Consequently, it was believed that if average grain size could be decreased even further to the nanometer length scale the yield strength would increase as well. However, experiments on many nanocrystalline materials demonstrated that if the grains reached a small enough size, the critical grain size which is typically around 10 nm (3.9×10^{-7} in), the yield strength would either remain constant or decrease with decreasing grains size. This phenomenon has been termed the reverse or inverse Hall–Petch relation. A number of different mechanisms have been proposed for this relation. As suggested by Carlton *et al.*, they fall into four categories: (1) dislocation-based, (2) diffusion-based, (3) grain-boundary shearing-based, (4) two-phase-based.

Other explanations that have been proposed to rationalize the apparent softening of metals with nanosized grains include poor sample quality and the suppression of dislocation pileups.

Many of the early measurements of a reverse Hall–Petch effect were likely the result of unrecognized pores in samples. The presence of voids in nanocrystalline metals would undoubtedly lead to their having weaker mechanical properties.

The pileup of dislocations at grain boundaries is a hallmark mechanism of the Hall–Petch relationship. Once grain sizes drop below the equilibrium distance between dislocations, though, this relationship should no longer be valid. Nevertheless, it is not entirely clear what exactly the dependency of yield stress should be on grain sizes below this point.

Grain Refinement

Grain refinement, also known as *inoculation*, is the set of techniques used to implement grain boundary strengthening in metallurgy. The specific techniques and corresponding mechanisms will vary based on what materials are being considered.

One method for controlling grain size in aluminum alloys is by introducing particles to serve as nucleants, such as Al–5%Ti. Grains will grow via heterogeneous nucleation; that is, for a given degree of undercooling beneath the melting temperature, aluminum particles in the melt will nucleate on the surface of the added particles. Grains will grow in the form of dendrites growing radially away from the surface of the nucleant. Solute particles can then be added (called grain refiners) which limit the growth of dendrites, leading to grain refinement. Al-Ti-B alloys are the most common grain refiner for Al alloys; however, novel refiners such as Al_3Sc have been suggested.

One common technique is to induce a very small fraction of the melt to solidify at a much higher temperature than the rest; this will generate seed crystals that act as a template when the rest of the material falls to its (lower) melting temperature and begins to solidify. Since a huge number of minuscule seed crystals are present, a nearly equal number of crystallites result, and the size of any one grain is limited.

Typical inoculants for various casting alloys	
Metal	Inoculant
Cast iron	FeSi, SiCa, graphite
Mg alloys	Zr, C
Cu alloys	Fe, Co, Zr
Al–Si alloys	P, Ti, B, Sc
Pb alloys	As, Te
Zn alloys	Ti
Ti alloys	Al–Ti intermetallics

Strengthening Mechanisms of Materials

Methods have been devised to modify the yield strength, ductility, and toughness of both crystalline and amorphous materials. These strengthening mechanisms give engineers the ability to tailor the mechanical properties of materials to suit a variety of different applications. For example, the favorable properties of steel result from interstitial incorporation of carbon into the iron lattice. Brass, a binary alloy of copper and zinc, has superior mechanical properties compared to its constituent metals due to solution strengthening. Work hardening (such as beating a red-hot piece of metal on anvil) has also been used for centuries by blacksmiths to introduce dislocations into materials, increasing their yield strengths.

What strengthening is

Plastic deformation occurs when large numbers of dislocations move and multiply so as to result in macroscopic deformation. In other words, it is the movement of dislocations in the material which allows for deformation. If we want to enhance a material's mechanical properties (i.e. increase the yield and tensile strength), we simply need to introduce a mechanism which prohibits the mobility of these dislocations. Whatever the mechanism may be, (work hardening, grain size reduction, etc.) they all hinder dislocation motion and render the material stronger than previously.

The stress required to cause dislocation motion is orders of magnitude lower than the theoretical stress required to shift an entire plane of atoms, so this mode of stress relief is energetically favorable. Hence, the hardness and strength (both yield and tensile) critically depend on the ease with which dislocations move. Pinning points, or locations in the crystal that oppose the motion of dislocations, can be introduced into the lattice to reduce dislocation mobility, thereby increasing mechanical strength. Dislocations may be pinned due to stress field interactions with other dislocations and solute particles, creating physical barriers from second phase precipitates forming along grain boundaries. There are four main strengthening mechanisms for metals, each is a method to prevent dislocation motion and propagation, or make it energetically unfavorable for the dislocation to move. For a material that has been strengthened, by some processing method, the amount of force required to start irreversible (plastic) deformation is greater than it was for the original material.

In amorphous materials such as polymers, amorphous ceramics (glass), and amorphous metals, the lack of long range order leads to yielding via mechanisms such as brittle fracture, crazing, and shear band formation. In these systems, strengthening mechanisms do not involve dislocations, but rather consist of modifications to the chemical structure and processing of the constituent material.

The strength of materials cannot infinitely increase. Each of the mechanisms explained below involves some trade-off by which other material properties are compromised in the process of strengthening.

Strengthening Mechanisms in Metals

Work Hardening

The primary species responsible for work hardening are dislocations. Dislocations interact with each other by generating stress fields in the material. The interaction between the stress fields of dislocations can impede dislocation motion by repulsive or attractive interactions. Additionally, if two dislocations cross, dislocation line entanglement occurs, causing the formation of a jog which opposes dislocation motion. These entanglements and jogs act as pinning points, which oppose dislocation motion. As both of these processes are more likely to occur when more dislocations are present, there is a correlation between dislocation density and yield strength,

$$\Delta \sigma_y = Gb\sqrt{\rho_\perp}$$

where G is the shear modulus, b is the Burgers vector, and $\rho_\perp$ is the dislocation density.

Increasing the dislocation density increases the yield strength which results in a higher shear stress required to move the dislocations. This process is easily observed while working a material (in metals cold working of process). Theoretically, the strength of a material with no dislocations will be extremely high ($\tau = G/2$) because plastic deformation would require the breaking of many bonds simultaneously. However, at moderate dislocation density values of around 10^7-10^9 dislocations/m^2, the material will exhibit a significantly lower mechanical strength. Analogously, it is easier to move a rubber rug across a surface by propagating a small ripple through it than by dragging the whole rug. At dislocation densities of 10^{14} dislocations/m^2 or higher, the strength of the material becomes high once again. Also, the dislocation density cannot be infinitely high, because then the material would lose its crystalline structure.

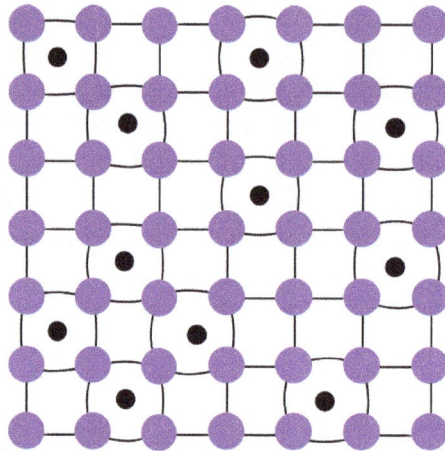

This is a schematic illustrating how the lattice is strained by the addition of interstitial solute. No-tice the strain in the lattice that the solute atoms cause. The interstitial solute could be carbon in iron for example. The carbon atoms in the interstitial sites of the lattice creates a stress field that impedes dislocation movement.

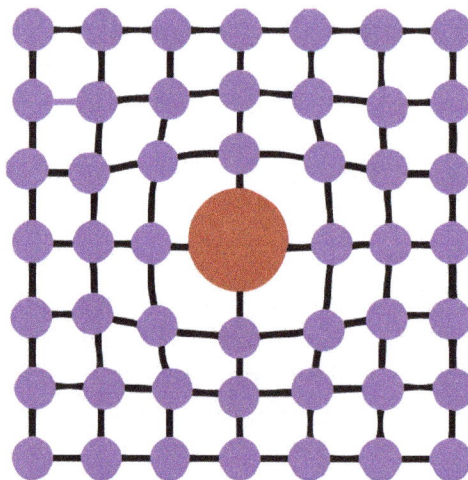

This is a schematic illustrating how the lattice is strained by the addition of substitutional solute. Notice the strain in the lattice that the solute atom causes.

Solid Solution Strengthening and Alloying

For this strengthening mechanism, solute atoms of one element are added to another, resulting in either substitutional or interstitial point defects in the crystal. The solute atoms cause lattice distortions that impede dislocation motion, increasing the yield stress of the material. Solute atoms have stress fields around them which can interact with those of dislocations. The presence of solute atoms impart compressive or tensile stresses to the lattice, depending on solute **size**, which interfere with nearby dislocations, causing the solute atoms to act as potential barriers \

The shear stress required to move dislocations in a material is:

$$\Delta\tau = Gb\sqrt{c}\epsilon^{3/2}$$

where c is the solute concentration and ϵ is the strain on the material caused by the solute.

Increasing the concentration of the solute atoms will increase the yield strength of a material, but there is a limit to the amount of solute that can be added, and one should look at the phase diagram for the material and the alloy to make sure that a second phase is not created.

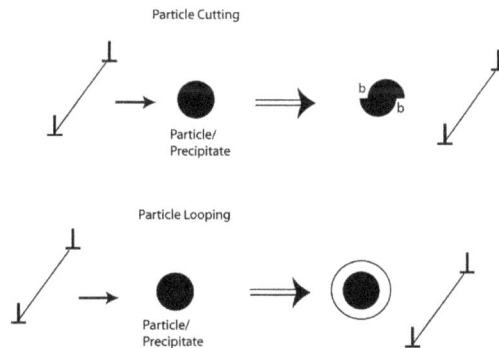

A schematic illustrating how the dislocations can interact with a particle. It can either cut through the particle or bow around the particle and create a dislocation loop as it moves over the particle.

In general, the solid solution strengthening depends on the concentration of the solute atoms, shear modulus of the solute atoms, size of solute atoms, valency of solute atoms (for ionic materials), and the symmetry of the solute stress field. The magnitude of strengthening is higher for non-symmetric stress fields because these solutes can interact with both edge and screw dislocations, whereas symmetric stress fields, which cause only volume change and not shape change, can only interact with edge dislocations.

Precipitation Hardening

In most binary systems, alloying above a concentration given by the phase diagram will cause the formation of a second phase. A second phase can also be created by mechanical or thermal treatments. The particles that compose the second phase precipitates act as pinning points in a similar manner to solutes, though the particles are not necessarily single atoms.

The dislocations in a material can interact with the precipitate atoms in one of two ways. If the precipitate atoms are small, the dislocations would cut through them. As a result, new surfaces of the particle would get exposed to the matrix and the particle-matrix interfacial energy would increase. For larger precipitate particles, looping or bowing of the dislocations would occur and result in dislocations getting longer. Hence, at a critical radius of about 5 nm, dislocations will preferably cut across the obstacle, while for a radius of 30 nm, the dislocations will readily bow or loop to overcome the obstacle.

The mathematical descriptions are as follows:

For particle bowing- $\Delta\tau = \dfrac{Gb}{L-2r}$

For particle cutting- $\Delta\tau = \dfrac{\gamma\pi r}{bL}$

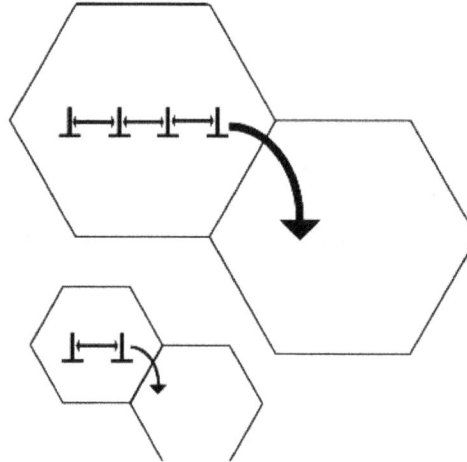

Figure : A schematic roughly illustrating the concept of dislocation pile up and how it effects the strength of the material. A material with larger grain size is able to have more dislocation to pile up leading to a bigger driving force for dislocations to move from one grain to another. Thus, less force need be applied to move a dislocation from a larger, than from a smaller grain, leading materials with smaller grains to exhibit higher yield stress.

Grain Boundary Strengthening

In a polycrystalline metal, grain size has a tremendous influence on the mechanical properties. Because grains usually have varying crystallographic orientations, grain boundaries arise. While undergoing deformation, slip motion will take place. Grain boundaries act as an impediment to dislocation motion for the following two reasons:

1. Dislocation must change its direction of motion due to the differing orientation of grains. 2. Discontinuity of slip planes from grain one to grain two.

The stress required to move a dislocation from one grain to another in order to plastically deform a material depends on the grain size. The average number of dislocations per grain decreases with average grain size. A lower number of dislocations per grain results in a lower dislocation 'pressure' building up at grain boundaries. This makes it more difficult for dislocations to move into adjacent grains. This relationship is the Hall-Petch relationship and can be mathematically described as follows:

$$\sigma_y = \sigma_{y,0} + \frac{k}{d^x},$$

where k is a constant, d is the average grain diameter and $\sigma_{y,0}$ is the original yield stress.

The fact that the yield strength increases with decreasing grain size is accompanied by the caveat that the grain size cannot be decreased infinitely. As the grain size decreases, more free volume is generated resulting in lattice mismatch. Below approximately 10 nm, the grain boundaries will tend to slide instead; a phenomenon known as grain-boundary sliding. If the grain size gets too small, it becomes more difficult to fit the dislocations in the grain and the stress required to move them is less. It was not possible to produce materials with grain sizes below 10 nm until recently, so the discovery that strength decreases below a critical grain size is still finding new applications.

Transformation Hardening

This method of hardening is used for steels.

High-strength steels generally fall into three basic categories, classified by the strengthening mechanism employed. 1- solid-solution-strengthened steels (rephos steels) 2- grain-refined steels or high strength low alloy steels (HSLA) 3- transformation-hardened steels

Transformation-hardened steels are the third type of high-strength steels.These steels use predominantly higher levels of C and Mn along with heat treatment to increase strength. The finished product will have a duplex micro-structure of ferrite with varying levels of degenerate marten-site. This allows for varying levels of strength. There are three basic types of transformation-hardened steels. These are dual-phase (DP), transformation-induced plasticity (TRIP),and martensitic steels.

The annealing process for dual -phase steels consists of first holding the steel in the alpha + gamma temperature region for a set period of time. During that time C and Mn diffuse into the austenite leaving a ferrite of greater purity. The steel is then quenched so that the Austen is transformed into martensite, and the ferrite remains on cooling. The steel is then subjected to a temper cycle to allow some level of marten-site decomposition. By controlling the amount of martensite in the steel, as well as the degree of temper, the strength level can be controlled. Depending on processing and chemistry, the strength level can range from 350 to 960 MPa.

TRIP steels also use C and Mn, along with heat treatment, in order to retain small amounts of Austen and bainite in a ferrite matrix. Thermal processing for TRIP steels again involves annealing the steel in the a + g region for a period of time sufficient to allow C and Mn to diffuse into austenite. The steel is then quenched to a point above the marten site start temperature and held there. This allows the formation of bainite, an austenite decomposition product. While at this temperature, more C is allowed to enrich the retained austenite. This, in turn, lowers the martensite start temperature to below room temperature. Upon final quenching a metastable austenite is retained in the predominantly ferrite matrix along with small amounts of bainite (and other forms of decomposed austenite). This combination of micro-structures has the added benefits of higher strengths and resistance to necking during forming. This offers great improvements in formability over other high-strength steels. Essentially, as the TRIP steel is being formed, it becomes much stronger. Tensile strengths of TRIP steels are in the range of 600-960 MPa.

Martensitic steels are also high in C and Mn. These are fully quenched to martensite during processing. The martensite structure is then tempered back to the appropriate strength level, adding toughness to the steel. Tensile strengths for these steels range as high as 1500 MPa.

Strengthening Mechanisms in Amorphous Materials

Polymer

Polymers fracture via breaking of inter- and intra molecular bonds; hence, the chemical structure of these materials plays a huge role in increasing strength. For polymers consisting of chains which easily slide past each other, chemical and physical cross linking can be used to increase rigidity and yield strength. In thermoset polymers (thermosetting plastic), disulfide bridges and other co-valent cross links give rise to a hard structure which can withstand very high temperatures. These cross-links are particularly helpful in improving tensile strength of materials which contain lots of free volume prone to crazing, typically glassy brittle polymers. In thermoplastic elastomer, phase separation of dissimilar monomer components leads to association of hard domains within a sea of soft phase, yielding a physical structure with increased strength and rigidity. If yielding occurs by chains sliding past each other (shear bands), the strength can also be increased by introducing kinks into the polymer chains via unsaturated carbon-carbon bonds.

Increasing the bulkiness of the monomer unit via incorporation of aryl rings is another strengthening mechanism. The anisotropy of the molecular structure means that these mechanisms are heavily dependent on the direction of applied stress. While aryl rings drastically increase rigidity along the direction of the chain, these materials may still be brittle in perpendicular directions. Macroscopic structure can be adjusted to compensate for this anisotropy. For example, the high strength of Kevlar arises from a stacked multilayer macrostructure where aromatic polymer layers are rotated with respect to their neighbors. When loaded oblique to the chain direction, ductile polymers with flexible linkages, such as oriented polyethylene, are highly prone to shear band formation, so macroscopic structures which place the load parallel to the draw direction would increase strength.

Mixing polymers is another method of increasing strength, particularly with materials that show crazing preceding brittle fracture such as atactic polystyrene (APS). For example, by forming a 50/50 mixture of APS with polyphenylene oxide (PPO), this embrittling tendency can be almost completely suppressed, substantially increasing the fracture strength.

Glass

Many silicate glasses are strong in compression but weak in tension. By introducing compression stress into the structure, the tensile strength of the material can be increased. This is typically done via two mechanisms: thermal treatment (tempering) or chemical bath (via ion exchange).

In tempered glasses, air jets are used to rapidly cool the top and bottom surfaces of a softened (hot) slab of glass. Since the surface cools quicker, there is more free volume at the surface than in the bulk melt. The core of the slab then pulls the surface inward, resulting in an internal compressive stress at the surface. This substantially increases the tensile strength of the material as tensile stresses exerted on the glass must now resolve the compressive stresses before yielding.

$$\sigma_{y=modified} = \sigma_{y,0} + \sigma_{compressive}$$

Alternately, in chemical treatment, a glass slab treated containing network formers and modifiers is submerged into a molten salt bath containing ions larger than those present in the modifier. Due to a concentration gradient of the ions, mass transport must take place. As the larger cation diffuses from the molten salt into the surface, it replaces the smaller ion from the modifier. The larger ion squeezing into surface introduces compressive stress in the glass's surface. A common example is treatment of sodium oxide modified silicate glass in molten potassium chloride.

Applications and Current Research

Strengthening of materials is useful in many applications. A primary application of strengthened materials is for construction. In order to have stronger buildings and bridges, one must have a strong frame that can support high tensile or compressive load and resist plastic deformation. The steel frame used to make the building should be as strong as possible so that it does not bend under the entire weight of the building. Polymeric roofing materials would also need to be strong so that the roof does not cave in when there is build-up of snow on the rooftop.

Research is also currently being done to increase the strength of metallic materials through the addition of polymer materials such as bonded carbon fiber reinforced polymer to (CFRP).

Molecular Dynamics Simulations

The use of computation simulations to model work hardening in materials allows for the direct observation of critical elements that rule the process of strengthening materials. The basic reasoning derives from the fact that, when examining plasticity and the movement of dislocations in materials, a focus on the atomistic level is many times not accounted for and the focus rests on the contiuum description of materials. Since the practice of tracking these atomistic effects in experiments and theorizing about them in textbooks cannot provide a full understanding of these interactions, many turn to molecular dynamics simulations to develop this understanding.

The simulations work by utilizing the known atomic interactions between any two atoms and the relationship $F = ma$, so that the dislocations moving through the material are ruled by simple mechanical actions and reactions of the atoms. The interatomic potential usually utilized to estimate these interactions is the Lennard – Jones 12:6 potential. Lennard – Jones is widely accepted because its experimental shortcomings are well-known. These interactions are simply scaled up to millions or billions of atoms in some cases to simulate materials more accurately.

Molecular dynamic simulations display the interactions based upon the governing equations provided above for the strengthening mechanisms. They provide an effective way to see these mechanisms in action outside the painstaking realm of direct observation during experiments.

Solid Solution Strengthening

Solid solution strengthening is a type of alloying that can be used to improve the strength of a pure metal. The technique works by adding atoms of one element (the alloying element) to the crystalline lattice of another element (the base metal), forming a solid solution. The local nonuniformity in the lattice due to the alloying element makes plastic deformation more difficult by impeding dislocation motion. In contrast, alloying beyond the solubility limit can form a second phase, leading to strengthening via other mechanisms (e.g. the precipitation of intermetallic compounds).

Types

Depending on the size of the alloying element, a substitutional solid solution or an interstitial solid solution can form. In both cases, the overall crystal structure is essentially unchanged.

Substitutional solid solution strengthening occurs when the solute atom is large enough that it can replace solvent atoms in their lattice positions. Some alloying elements are only soluble in small amounts, whereas some solvent and solute pairs form a solution over the whole range of binary compositions. Generally, higher solubility is seen when solvent and solute atoms are similar in atomic size (15% according to the Hume-Rothery rules) and adopt the same crystal structure in their pure form. Examples of completely miscible binary systems are Cu-Ni and the Ag-Au face-centered cubic (FCC) binary systems, and the Mo-W body-centered cubic (BCC) binary system.

Interstitial solid solutions form when the solute atom is small enough to fit at interstitial sites between the solvent atoms. The atoms crowd into the interstitial sites, causing the bonds of the solvent atoms to compress and thus deform. Elements commonly used to form interstitial solid solutions include H, Li, Na, N, C, and O. Carbon in iron (steel) is one example of interstitial solid solution.

Mechanism

The strength of a material is dependent on how easily dislocations in its crystal lattice can be propagated. These dislocations create stress fields within the material depending on their character. When solute atoms are introduced, local stress fields are formed that interact with those of the dislocations, impeding their motion and causing an increase in the yield stress of the material, which means an increase in strength of the material. This gain is a result of both lattice distortion and the modulus effect.

When solute and solvent atoms differ in size, local stress fields are created that can attract or repel dislocations in their vicinity. This is known as the size effect. By relieving tensile or compressive strain in the lattice, the solute size mismatch can put the dislocation in a lower energy state. In substitutional solid solutions, these stress fields are spherically symmetric, meaning they have no shear stress component. As such, substitutional solute atoms do not interact with the shear stress fields characteristic of screw dislocations. Conversely, in interstitial solid solutions, solute atoms cause a tetragonal distortion, generating a shear field that can interact with edge, screw, and mixed dislocations. The attraction or repulsion of the dislocation centers to the solute particles increase the stress it takes to propagate the dislocation in any other direction. Increasing the applied stress necessary to move the dislocation increases the yield strength of the material.

The energy density of a dislocation is dependent on its Burgers vector as well as the modulus of the local atoms. When the modulus of solute atoms differs from that of the host element, the local energy around the dislocation is changed, increasing the amount of force necessary to move past this energy well. This is known as the modulus effect.

Surface carburizing, or case hardening, is one example of solid solution strengthening in which the density of solute carbon atoms is increased close to the surface of the steel, resulting in a gradient of carbon atoms throughout the material. This provides superior mechanical properties to the surface of the steel.

Governing Equations

Solid solution strengthening increases yield strength of the material by increasing the stress to move dislocations:

$$\Delta \tau = Gb\epsilon^{\frac{3}{2}}\sqrt{c}$$

where c is the concentration of the solute atoms, G is the shear modulus, b is the magnitude of the Burger's vector, and $\dot{o}$ is the lattice strain due to the solute. This is composed of two terms, one describing lattice distortion and the other local modulus change.

$\epsilon = |\epsilon_G - \beta\epsilon_a|$ Here, ϵ_G the term that captures the local modulus change, β a constant dependent on the solute atoms and ϵ_a is the lattice distortion term.

The lattice distortion term can be described as:

$$\epsilon_a = \frac{\Delta a}{a\Delta c}, \text{ where } a \text{ is the lattice parameter of the material.}$$

Meanwhile, the local modulus change is captured in the following expression:

$$\epsilon_G = \frac{\Delta G}{G\Delta c}, \text{ where } G \text{ is shear modulus of the solute material,}$$

Implications

In order to achieve noticeable material strengthening via solution strengthening, one should alloy with solutes of higher shear modulus, hence increasing the local shear modulus in the material. In addition, one should alloy with elements of different equilibrium lattice constants. The greater the difference in lattice parameter, the higher the local stress fields introduced by alloying. Alloying with elements of higher shear modulus or of very different lattice parameters will increase the stiffness and introduce local stress fields respectively. In either case, the dislocation propagation will be hindered at these sites, impeding plasticity and increasing yield strength proportionally with solute concentration.

Solid solution strengthening depends on:

- Concentration of solute atoms
- Shear modulus of solute atoms
- Size of solute atoms
- Valency of solute atoms (for ionic materials)

Nevertheless, one should not add so much solute as to precipitate a new phase. This occurs if the concentration of the solute reaches a certain critical point given by the binary system phase diagram. This critical concentration therefore puts a limit to the amount of solid solution strengthening that can be achieved with a given material.

References

- Stefanescu, Doru Michael (2002), Science and engineering of casting solidification, Springer, p. 265, ISBN 978-0-306-46750-9.

- Pelleg, Joshua (2013). Mechanical Properties of Materials. New York: Springer. pp. 236–239. ISBN 978-94-007-4341-0.

Properties of Ceramics

The properties of ceramics are high-temperature superconductivity, piezoelectricity, compressive strength, indentation hardness and stiffness. High-temperature superconductivity is the manner in which materials behave at high temperatures. The properties elucidated in this section are of vital importance and provides a better understanding of ceramics.

High-temperature Superconductivity

High-temperature superconductors (abbreviated high-T_c or HTS) are materials that behave as superconductors at unusually high temperatures. The first high-T_c superconductor was discovered in 1986 by IBM researchers Georg Bednorz and K. Alex Müller, who were awarded the 1987 Nobel Prize in Physics "for their important break-through in the discovery of superconductivity in ceramic materials".

A small sample of the high-temperature superconductor BSCCO-2223.

Whereas "ordinary" or metallic superconductors usually have transition temperatures (temperatures below which they are superconductive) below 30 K (−243.2 °C), and must be cooled using liquid helium in order to achieve superconductivity, HTS have been observed with transition temperatures as high as 138 K (−135 °C), and can be cooled to superconductivity using liquid nitrogen. Until 2008, only certain compounds of copper and oxygen (so-called "cuprates") were believed to have HTS properties, and the term high-temperature superconductor was used interchangeably with cuprate superconductor for compounds such as bismuth strontium calcium copper oxide (BSCCO) and yttrium barium copper oxide (YBCO). Several iron-based compounds (the iron pnictides) are now known to be superconducting at high temperatures.

In 2015, hydrogen sulfide (H_2S) under extremely high pressure (around 150 gigapascals) was found to undergo superconducting transition near 203 K (-70 °C), the highest temperature superconductor known to date.

History

The phenomenon of superconductivity was discovered by Kamerlingh Onnes in 1911, in metallic mercury below 4 K (-269.15 °C). For seventy-five years after that, researchers attempted to observe superconductivity at higher and higher temperatures. In the late 1970s, superconductivity was observed in certain metal oxides at temperatures as high as 13 K (-260.1 °C), which were much higher than those for elemental metals. In 1986, J. Georg Bednorz and K. Alex Müller, working at the IBM research lab near Zurich, Switzerland were exploring a new class of ceramics for superconductivity. Bednorz encountered a barium-doped compound of lanthanum and copper oxide whose resistance dropped down to zero at a temperature around 35 K (-238.2 °C). Their results were soon confirmed by many groups, notably Paul Chu at the University of Houston and Shoji Tanaka at the University of Tokyo.

Shortly after, P. W. Anderson, at Princeton University came up with the first theoretical description of these materials, using the resonating valence bond theory, but a full understanding of these materials is still developing today. These superconductors are now known to possess a d-wave pair symmetry. The first proposal that high-temperature cuprate superconductivity involves d-wave pairing was made in 1987 by Bickers, Scalapino and Scalettar, followed by three subsequent theories in 1988 by Inui, Doniach, Hirschfeld and Ruckenstein, using spin-fluctuation theory, and by Gros, Poilblanc, Rice and Zhang, and by Kotliar and Liu identifying d-wave pairing as a natural consequence of the RVB theory. The confirmation of the d-wave nature of the cuprate superconductors was made by a variety of experiments, including the direct observation of the d-wave nodes in the excitation spectrum through Angle Resolved Photoemission Spectroscopy, the observation of a half-integer flux in tunneling experiments, and indirectly from the temperature dependence of the penetration depth, specific heat and thermal conductivity.

The superconductor with the highest transition temperature that has been confirmed by multiple independent research groups (a prerequisite to be called a discovery, verified by peer review) is mercury barium calcium copper oxide ($HgBa_2Ca_2Cu_3O_8$) at around 133 K.

After more than twenty years of intensive research, the origin of high-temperature superconductivity is still not clear, but it seems that instead of *electron-phonon* attraction mechanisms, as in conventional superconductivity, one is dealing with genuine *electronic* mechanisms (e.g. by antiferromagnetic correlations), and instead of conventional, purely s-wave pairing, more exotic pairing symmetries are thought to be involved (d-wave in the case of the cuprates; primarily extended s-wave, but occasionally d-wave, in the case of the iron-based superconductors). One goal of all this research is room-temperature superconductivity. In 2014, evidence showing that fractional particles can happen in quasi two-dimensional magnetic materials, was found by EPFL scientists lending support for Anderson's theory of high-temperature superconductivity.

Crystal Structures of High-temperature Ceramic Superconductors

The structure of high-T_c copper oxide or cuprate superconductors are often closely related to perovskite structure, and the structure of these compounds has been described as a distorted, oxygen deficient multi-layered perovskite structure. One of the properties of the crystal structure of oxide superconductors is an alternating multi-layer of CuO_2 planes with superconductivity taking place between these layers. The more layers of CuO_2, the higher T_c. This structure causes a large anisotropy in normal conducting and superconducting properties, since electrical currents are carried by holes induced in the oxygen sites of the CuO_2 sheets. The electrical conduction is highly anisotropic, with a much higher conductivity parallel to the CuO_2 plane than in the perpendicular direction. Generally, critical temperatures depend on the chemical compositions, cations substitutions and oxygen content. They can be classified as superstripes; i.e., particular realizations of superlattices at atomic limit made of superconducting atomic layers, wires, dots separated by spacer layers, that gives multiband and multigap superconductivity.

YBaCuO Superconductors

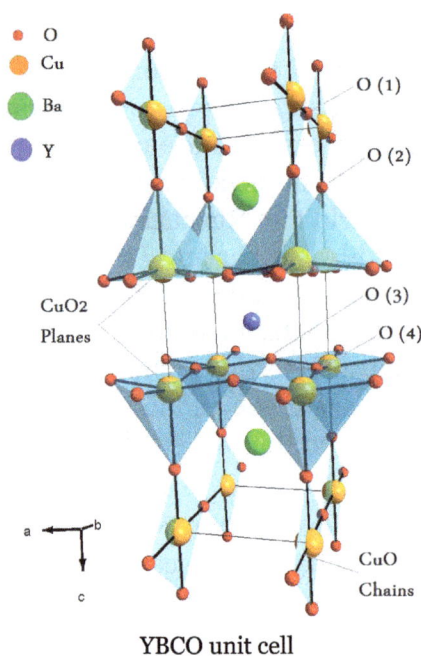

YBCO unit cell

The first superconductor found with $T_c > 77$ K (liquid nitrogen boiling point) is yttrium barium copper oxide ($YBa_2Cu_3O_{7-x}$); the proportions of the three different metals in the $YBa_2Cu_3O_7$ superconductor are in the mole ratio of 1 to 2 to 3 for yttrium to barium to copper, respectively. Thus, this particular superconductor is often referred to as the 123 superconductor.

The unit cell of $YBa_2Cu_3O_7$ consists of three pseudocubic elementary perovskite unit cells. Each perovskite unit cell contains a Y or Ba atom at the center: Ba in the bottom unit cell, Y in the middle one, and Ba in the top unit cell. Thus, Y and Ba are stacked in the sequence [Ba–Y–Ba] along the c-axis. All corner sites of the unit cell are occupied by Cu, which has two different coordinations, Cu(1) and Cu(2), with respect to oxygen. There are four possible crystallographic sites for oxygen: O(1), O(2), O(3) and O(4). The coordination polyhedra of Y and Ba with respect to oxygen are different. The tripling of the perovskite unit cell leads to nine oxygen atoms, whereas $YBa_2Cu_3O_7$ has seven oxygen atoms and, therefore, is referred to as an oxygen-deficient perovskite structure. The structure has a stacking of different layers: $(CuO)(BaO)(CuO_2)(Y)(CuO_2)(BaO)(CuO)$. One of the key feature of the unit cell of $YBa_2Cu_3O_{7-x}$ (YBCO) is the presence of two layers of CuO_2. The role of the Y plane is to serve as a spacer between two CuO_2 planes. In YBCO, the Cu–O chains are known to play an important role for superconductivity. T_c is maximal near 92 K when $x \approx 0.15$ and the structure is orthorhombic. Superconductivity disappears at $x \approx 0.6$, where the structural transformation of YBCO occurs from orthorhombic to tetragonal.

Bi-, Tl- and Hg-based high-T_c Superconductors

The crystal structure of Bi-, Tl- and Hg-based high-T_c superconductors are very similar. Like YBCO, the perovskite-type feature and the presence of CuO_2 layers also exist in these superconductors. However, unlike YBCO, Cu–O chains are not present in these superconductors. The YBCO superconductor has an orthorhombic structure, whereas the other high-T_c superconductors have a tetragonal structure.

The Bi–Sr–Ca–Cu–O system has three superconducting phases forming a homologous series as $Bi_2Sr_2Ca_{n-1}Cu_nO_{4+2n+x}$ (n = 1, 2 and 3). These three phases are Bi-2201, Bi-2212 and Bi-2223, having transition temperatures of 20, 85 and 110 K, respectively, where the numbering system represent number of atoms for Bi, Sr, Ca and Cu respectively. The two phases have a tetragonal structure which consists of two sheared crystallographic unit cells. The unit cell of these phases has double Bi–O planes which are stacked in a way that the Bi atom of one plane sits below the oxygen atom of the next consecutive plane. The Ca atom forms a layer within the interior of the CuO_2 layers in both Bi-2212 and Bi-2223; there is no Ca layer in the Bi-2201 phase. The three phases differ with each other in the number of CuO_2 planes; Bi-2201, Bi-2212 and Bi-2223 phases have one, two and three CuO_2 planes, respectively. The c axis lattice constants of these phases increases with the number of CuO_2 planes. The coordination of the Cu atom is different in the three phases. The Cu atom forms an octahedral coordination with respect to oxygen atoms in the 2201 phase, whereas in 2212, the Cu atom is surrounded by five oxygen atoms in a pyramidal arrangement. In the 2223 structure, Cu has two coordinations with respect to oxygen: one Cu atom is bonded with four oxygen atoms in square planar configuration and another Cu atom is coordinated with five oxygen atoms in a pyramidal arrangement.

Tl–Ba–Ca–Cu–O superconductor: The first series of the Tl-based superconductor containing one Tl–O layer has the general formula $TlBa_2Ca_{n-1}Cu_nO_{2n+3}$, whereas the second series containing two Tl–O layers has a formula of $Tl_2Ba_2Ca_{n-1}Cu_nO_{2n+4}$ with n = 1, 2 and 3. In the structure of $Tl_2Ba_2CuO_6$ (Tl-2201), there is one CuO_2 layer with the stacking sequence (Tl–O) (Tl–O) (Ba–O) (Cu–O) (Ba–O) (Tl–O) (Tl–O). In $Tl_2Ba_2CaCu_2O_8$ (Tl-2212), there are two Cu–O layers with a Ca layer in between. Similar to the $Tl_2Ba_2CuO_6$ structure, Tl–O layers are present outside the Ba–O layers. In $Tl_2Ba_2Ca_2Cu_3O_{10}$ (Tl-2223), there are three CuO2 layers enclosing Ca layers between each of these. In Tl-based superconductors, T_c is found to increase with the increase in CuO_2 layers. However, the value of T_c decreases after four CuO_2 layers in $TlBa_2Ca_{n-1}Cu_nO_{2n+3}$, and in the $Tl_2Ba_2Ca_{n-1}Cu_nO_{2n+4}$ compound, it decreases after three CuO_2 layers.

Hg–Ba–Ca–Cu–O superconductor: The crystal structure of $HgBa_2CuO_4$ (Hg-1201), $HgBa_2CaCu_2O_6$ (Hg-1212) and $HgBa_2Ca_2Cu_3O_8$ (Hg-1223) is similar to that of Tl-1201, Tl-1212 and Tl-1223, with Hg in place of Tl. It is noteworthy that the T_c of the Hg compound (Hg-1201) containing one CuO_2 layer is much larger as compared to the one-CuO_2-layer compound of thallium (Tl-1201). In the Hg-based superconductor, T_c is also found to increase as the CuO_2 layer increases. For Hg-1201, Hg-1212 and Hg-1223, the values of T_c are 94, 128 and the record value at ambient pressure 134 K, respectively, as shown in table below. The observation that the T_c of Hg-1223 increases to 153 K under high pressure indicates that the T_c of this compound is very sensitive to the structure of the compound.

Critical temperature (T_c), crystal structure and lattice constants of some high-T_c superconductors				
Formula	**Notation**	**T_c (K)**	**No. of Cu-O planes in unit cell**	**Crystal structure**
$YBa_2Cu_3O_7$	123	92	2	Orthorhombic
$Bi_2Sr_2CuO_6$	Bi-2201	20	1	Tetragonal
$Bi_2Sr_2CaCu_2O_8$	Bi-2212	85	2	Tetragonal
$Bi_2Sr_2Ca_2Cu_3O_{10}$	Bi-2223	110	3	Tetragonal
$Tl_2Ba_2CuO_6$	Tl-2201	80	1	Tetragonal
$Tl_2Ba_2CaCu_2O_8$	Tl-2212	108	2	Tetragonal
$Tl_2Ba_2Ca_2Cu_3O_{10}$	Tl-2223	125	3	Tetragonal
$TlBa_2Ca_3Cu_4O_{11}$	Tl-1234	122	4	Tetragonal
$HgBa_2CuO_4$	Hg-1201	94	1	Tetragonal
$HgBa_2CaCu_2O_6$	Hg-1212	128	2	Tetragonal
$HgBa_2Ca_2Cu_3O_8$	Hg-1223	134	3	Tetragonal

Preparation of high-T_c Superconductors

The simplest method for preparing high-T_c superconductors is a solid-state thermochemical reaction involving mixing, calcination and sintering. The appropriate amounts of precursor powders, usually oxides and carbonates, are mixed thoroughly using a Ball mill. Solution chemistry processes such as coprecipitation, freeze-drying and sol-gel methods are alternative ways for preparing a homogeneous mixture. These powders are calcined in the temperature range from 800 °C to 950 °C for several hours. The powders are cooled, reground and calcined again. This process is repeated several times to get homogeneous material. The powders are subsequently compacted to pellets and sintered. The sintering environment such as temperature, annealing time, atmosphere and cooling rate play a very important role in getting good high-T_c superconducting materials. The $YBa_2Cu_3O_{7-x}$ compound is prepared by calcination and sintering of a homogeneous mixture of Y_2O_3, $BaCO_3$ and CuO in the appropriate atomic ratio. Calcination is done at 900–950 °C, whereas sintering is done at 950 °C in an oxygen atmosphere. The oxygen stoichiometry in this material is very crucial for obtaining a superconducting $YBa_2Cu_3O_{7-x}$ compound. At the time of sintering, the semiconducting tetragonal $YBa_2Cu_3O_6$ compound is formed, which, on slow cooling in oxygen atmosphere, turns into superconducting $YBa_2Cu_3O_{7-x}$. The uptake and loss of oxygen are reversible in $YBa_2Cu_3O_{7-x}$. A fully oxygenated orthorhombic $YBa_2Cu_3O_{7-x}$ sample can be transformed into tetragonal $YBa_2Cu_3O_6$ by heating in a vacuum at temperature above 700 °C.

The preparation of Bi-, Tl- and Hg-based high-T_c superconductors is difficult compared to YBCO. Problems in these superconductors arise because of the existence of three or more phases having a similar layered structure. Thus, syntactic intergrowth and defects such as stacking faults occur during synthesis and it becomes difficult to isolate a single superconducting phase. For Bi–Sr–Ca–Cu–O, it is relatively simple to prepare the Bi-2212 ($T_c \approx 85$ K) phase, whereas it is very difficult to prepare a single phase of Bi-2223 ($T_c \approx 110$ K). The Bi-2212 phase appears only after few hours of sintering at 860–870 °C, but the larger fraction of the Bi-2223 phase is formed after a long reaction time of more than a week at 870 °C. Although the substitution of Pb in the Bi–Sr–Ca–Cu–O compound has been found to promote the growth of the high-T_c phase, a long sintering time is still required.

Properties

"High-temperature" has two common definitions in the context of superconductivity:

1. Above the temperature of 30 K that had historically been taken as the upper limit allowed by BCS theory(1957). This is also above the 1973 record of 23 K that had lasted until copper-oxide materials were discovered in 1986.

2. Having a transition temperature that is a larger fraction of the Fermi temperature than for conventional superconductors such as elemental mercury or lead. This definition encompasses a wider variety of unconventional superconductors and is used in the context of theoretical models.

The label high-T_c may be reserved by some authors for materials with critical temperature greater than the boiling point of liquid nitrogen (77 K or −196 °C). However, a number of materials – including the original discovery and recently discovered pnictide superconductors – had critical temperatures below 77 K but are commonly referred to in publication as being in the high-T_c class.

Technological applications could benefit from both the higher critical temperature being above the boiling point of liquid nitrogen and also the higher critical magnetic field (and critical current density) at which superconductivity is destroyed. In magnet applications, the high critical magnetic field may prove more valuable than the high T_c itself. Some cuprates have an upper critical field of about 100 tesla. However, cuprate materials are brittle ceramics which are expensive to manufacture and not easily turned into wires or other useful shapes. Also, high-temperature superconductors do not form large, continuous superconducting domains, but only clusters of microdomains within which superconducting occurs. They are therefore unsuitable for applications requiring actual superconducted currents, such as magnets for magnetic resonance spectrometers.

After two decades of intense experimental and theoretical research, with over 100,000 published papers on the subject, several common features in the properties of high-temperature superconductors have been identified. As of 2011, no widely accepted theory explains their properties. Relative to conventional superconductors, such as elemental mercury or lead that are adequately explained by the BCS theory, cuprate superconductors (and other unconventional superconductors) remain distinctive. There also has been much debate as to high-temperature superconductivity coexisting with magnetic ordering in YBCO, iron-based superconductors, several ruthenocuprates and other exotic superconductors, and the search continues for other families of materials. HTS are Type-II superconductors, which allow magnetic fields to penetrate their interior in quantized units of flux, meaning that much higher magnetic fields are required to suppress superconductivity. The layered structure also gives a directional dependence to the magnetic field response.

Cuprates

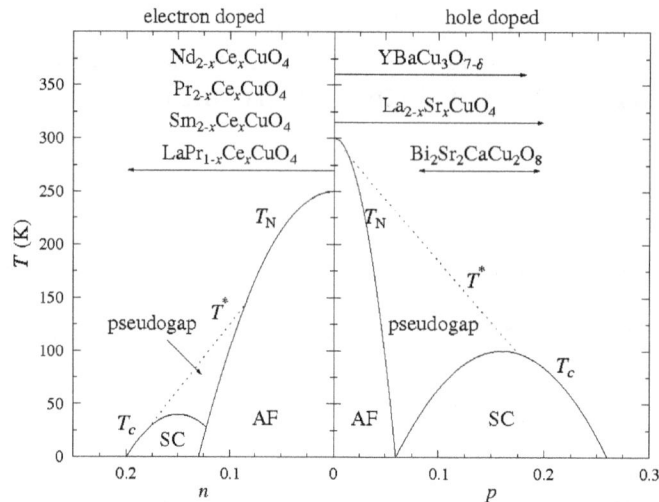

Simplified doping dependent phase diagram of cuprate superconductors for both electron (n) and hole (p) doping. The phases shown are the antiferromagnetic (AF) phase close to zero doping, the superconducting phase around optimal doping, and the pseudogap phase. Doping ranges possible for some common compounds are also shown.

Cuprate superconductors are generally considered to be quasi-two-dimensional materials with their superconducting properties determined by electrons moving within weakly coupled copper-oxide (CuO_2) layers. Neighbouring layers containing ions such as lanthanum, barium, strontium, or other atoms act to stabilize the structure and dope electrons or holes onto the copper-oxide layers. The undoped "parent" or "mother" compounds are Mott insulators with long-range antiferromagnetic order at low enough temperature. Single band models are generally considered to be sufficient to describe the electronic properties.

The cuprate superconductors adopt a perovskite structure. The copper-oxide planes are checkerboard lattices with squares of O^{2-} ions with a Cu^{2+} ion at the centre of each square. The unit cell is rotated by 45° from these squares. Chemical formulae of superconducting materials generally contain fractional numbers to describe the doping required for superconductivity. There are several families of cuprate superconductors and they can be categorized by the elements they contain and the number of adjacent copper-oxide layers in each superconducting block. For example, YBCO and BSCCO can alternatively be referred to as Y123 and Bi2201/Bi2212/Bi2223 depending on the number of layers in each superconducting block (n). The superconducting transition temperature has been found to peak at an optimal doping value ($p = 0.16$) and an optimal number of layers in each superconducting block, typically $n = 3$.

Possible mechanisms for superconductivity in the cuprates are still the subject of considerable debate and further research. Certain aspects common to all materials have been identified. Similarities between the antiferromagnetic low-temperature state of the undoped materials and the superconducting state that emerges upon doping, primarily the $d_{x^2-y^2}$ orbital state of the Cu^{2+} ions, suggest that electron-electron interactions are more significant than electron-phonon interactions in cuprates – making the superconductivity unconventional. Recent work on the Fermi surface has shown that nesting occurs at four points in the antiferromagnetic Brillouin zone where spin waves exist and that the superconducting energy gap is larger at these points. The weak isotope effects observed for most cuprates contrast with conventional superconductors that are well described by BCS theory.

Similarities and differences in the properties of hole-doped and electron doped cuprates:

- Presence of a pseudogap phase up to at least optimal doping.

- Different trends in the Uemura plot relating transition temperature to the superfluid density. The inverse square of the London penetration depth appears to be proportional to the critical temperature for a large number of underdoped cuprate superconductors, but the constant of proportionality is different for hole- and electron-doped cuprates. The linear trend implies that the physics of these materials is strongly two-dimensional.

- Universal hourglass-shaped feature in the spin excitations of cuprates measured using inelastic neutron diffraction.

- Nernst effect evident in both the superconducting and pseudogap phases.

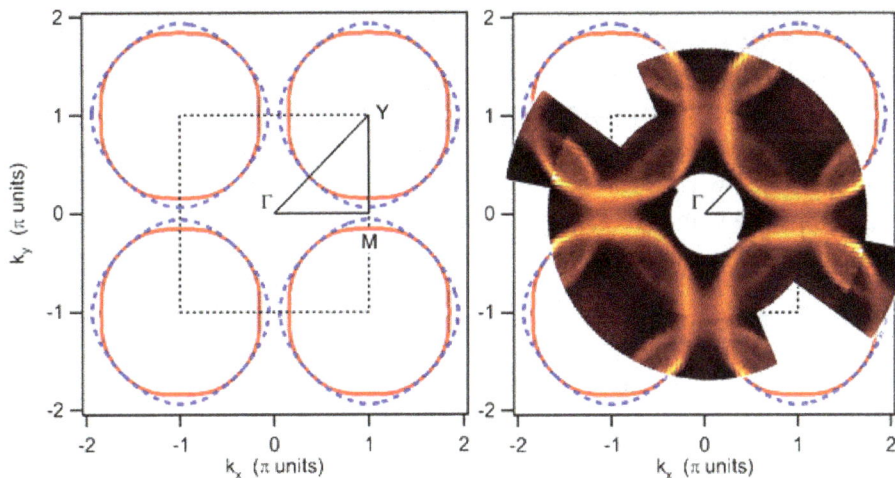

Fig. The Fermi surface of bi-layer BSCCO, calculated (left) and measured by ARPES (right). The dashed rectangle represents the first Brillouin zone.

The electronic structure of superconducting cuprates is highly anisotropic. Therefore, the Fermi surface of HTSC is very close to the Fermi surface of the doped CuO_2 plane (or multi-planes, in case of multi-layer cuprates) and can be presented on the 2D reciprocal space (or momentum space) of the CuO_2 lattice. The typical Fermi surface within the first CuO_2 Brillouin zone is sketched in Fig. (left). It can be derived from the band structure calculations or measured by angle resolved photoemission spectroscopy (ARPES). Fig. (right) shows the Fermi surface of BSCCO measured by ARPES. In a wide range of charge carrier concentration (doping level), in which the hole-doped HTSC are superconducting, the Fermi surface is hole-like (i.e. open, as shown in Fig.). This results in an inherent in-plane anisotropy of the electronic properties of HTSC.

Iron-based Superconductors

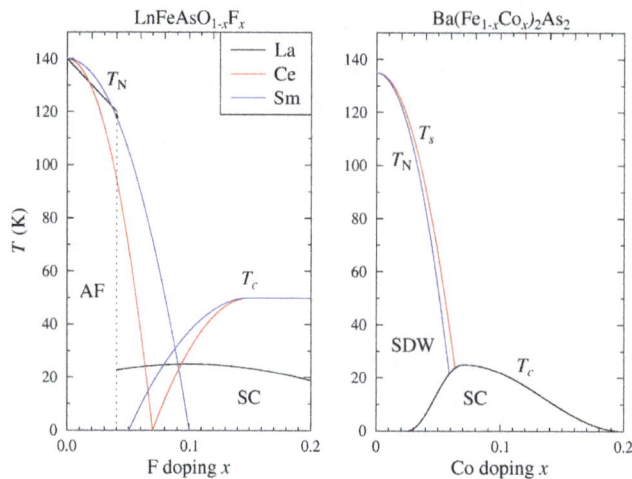

Simplified doping dependent phase diagrams of iron-based superconductors for both Ln-1111 and Ba-122 materials. The phases shown are the antiferromagnetic/spin density wave (AF/SDW) phase close to zero doping and the superconducting phase around optimal doping. The Ln-1111 phase diagrams for La and Sm were determined using muon spin spectroscopy, the phase diagram for Ce was determined using neutron diffraction. The Ba-122 phase diagram is based on.

Iron-based superconductors contain layers of iron and a pnictogen—such as arsenic or phosphorus—or a chalcogen. This is currently the family with the second highest critical temperature, behind the cuprates. Interest in their superconducting properties began in 2006 with the discovery of superconductivity in LaFePO at 4 K and gained much greater attention in 2008 after the analogous material LaFeAs(O,F) was found to superconduct at up to 43 K under pressure. The highest critical temperatures in the iron-based superconductor family exist in thin films of FeSe, where a critical temperature in excess of 100 K has recently been reported.

Since the original discoveries several families of iron-based superconductors have emerged:

- LnFeAs(O,F) or LnFeAsO$_{1-x}$ (Ln = lanthanide) with T_c up to 56 K, referred to as 1111 materials. A fluoride variant of these materials was subsequently found with similar T_c values.

- (Ba,K)Fe$_2$As$_2$ and related materials with pairs of iron-arsenide layers, referred to as 122 compounds. T_c values range up to 38 K. These materials also superconduct when iron is replaced with cobalt

- LiFeAs and NaFeAs with T_c up to around 20 K. These materials superconduct close to stoichiometric composition and are referred to as 111 compounds.

- FeSe with small off-stoichiometry or tellurium doping.

Most undoped iron-based superconductors show a tetragonal-orthorhombic structural phase transition followed at lower temperature by magnetic ordering, similar to the cuprate superconductors. However, they are poor metals rather than Mott insulators and have five bands at the Fermi surface rather than one. The phase diagram emerging as the iron-arsenide layers are doped is remarkably similar, with the superconducting phase close to or overlapping the magnetic phase. Strong evidence that the T_c value varies with the As-Fe-As bond angles has already emerged and shows that the optimal T_c value is obtained with undistorted $FeAs_4$ tetrahedra. The symmetry of the pairing wavefunction is still widely debated, but an extended s-wave scenario is currently favoured.

Hydrogen Sulfide

At pressures above 90 GPa (Gigapascal), hydrogen sulfide becomes a metallic conductor of electricity. When cooled below a critical temperature this high-pressure phase exhibits superconductivity. The critical temperature increases with pressure, ranging from 23 K at 100 GPa to 150 K at 200 GPa. If hydrogen sulfide is pressurized at higher temperatures, then cooled, the critical temperature reaches 203 K (−70 °C), the highest accepted superconducting critical temperature as of 2015. It has been predicted that by substituting a small part of sulfur with phosphorus and using even higher pressures it may be possible to raise the critical temperature to above 0 °C (273 K) and achieve room-temperature superconductivity.

Other Materials Sometimes Referred to as High-temperature Superconductors

Magnesium diboride is occasionally referred to as a high-temperature superconductor because its T_c value of 39 K is above that historically expected for BCS superconductors. However, it is more generally regarded as the highest-T_c conventional superconductor, the increased T_c resulting from two separate bands being present at the Fermi level.

Fulleride superconductors where alkali-metal atoms are intercalated into C_{60} molecules show superconductivity at temperatures of up to 38 K for Cs_3C_{60}.

Some organic superconductors and heavy fermion compounds are considered to be high-temperature superconductors because of their high T_c values relative to their Fermi energy, despite the T_c values being lower than for many conventional superconductors. This description may relate better to common aspects of the superconducting mechanism than the superconducting properties.

Metallic Hydrogen

Theoretical work by Neil Ashcroft in 1968 predicted that solid metallic hydrogen at extremely high pressure should become superconducting at approximately room-temperature because of its extremely high speed of sound and expected strong coupling between the conduction electrons and the lattice vibrations. This prediction is yet to be experimentally verified.

Magnetic Properties

All known high-T_c superconductors are Type-II superconductors. In contrast to Type-I superconductors, which expel all magnetic fields due to the Meissner effect, Type-II superconductors allow magnetic fields to penetrate their interior in quantized units of flux, creating "holes" or "tubes" of normal metallic regions in the superconducting bulk called vortices. Consequently, high-T_c superconductors can sustain much higher magnetic fields.

Ongoing Research

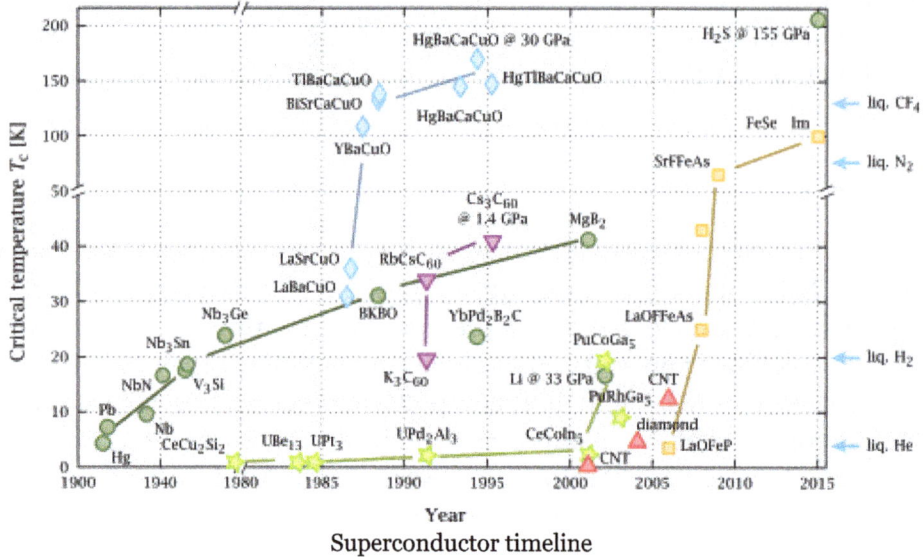

Superconductor timeline

The question of how superconductivity arises in high-temperature superconductors is one of the major unsolved problems of theoretical condensed matter physics. The mechanism that causes the electrons in these crystals to form pairs is not known. Despite intensive research and many promising leads, an explanation has so far eluded scientists. One reason for this is that the materials in question are generally very complex, multi-layered crystals (for example, BSCCO), making theoretical modelling difficult.

Improving the quality and variety of samples also gives rise to considerable research, both with the aim of improved characterisation of the physical properties of existing compounds, and synthesizing new materials, often with the hope of increasing T_c. Technological research focuses on making HTS materials in sufficient quantities to make their use economically viable and optimizing their properties in relation to applications.

Possible Mechanism

There have been two representative theories for high-temperature or unconventional superconductivity. Firstly, weak coupling theory suggests superconductivity emerges from antiferromagnetic spin fluctuations in a doped system. According to this theory, the pairing wave function of the cuprate HTS should have a $d_{x^2-y^2}$ symmetry. Thus, determining whether the pairing wave function has d-wave symmetry is essential to test the spin fluctuation mechanism. That is, if the HTS order parameter (pairing wave function) does not have d-wave symmetry, then a pairing mechanism related to spin fluctuations can be ruled out. (Similar arguments can be made for iron-based superconductors but the different material properties allow a different pairing symmetry.) Secondly, there was the interlayer coupling model, according to which a layered structure consisting of BCS-type (s-wave symmetry) superconductors can enhance the superconductivity by itself. By introducing an additional tunnelling interaction between each layer, this model successfully explained the anisotropic symmetry of the order parameter as well as the emergence of the HTS. Thus, in order to solve this unsettled problem, there have been numerous experiments such as photoemission spectroscopy, NMR, specific heat measurements, etc. Up to date the results were ambiguous, some reports supported the *d* symmetry for the HTS whereas others supported the *s* symmetry. This muddy situation possibly originated from the indirect nature of the experimental evidence, as well as experimental issues such as sample quality, impurity scattering, twinning, etc.

Junction Experiment Supporting the d Symmetry

The Meissner effect or a magnet levitating above a superconductor (cooled by liquid nitrogen)

An experiment based on flux quantization of a three-grain ring of $YBa_2Cu_3O_7$ (YBCO) was proposed to test the symmetry of the order parameter in the HTS. The symmetry of the order parameter could best be probed at the junction interface as the Cooper pairs tunnel across a Josephson junction or weak link. It was expected that a half-integer flux, that is, a spontaneous magnetization could only occur for a junction of *d* symmetry superconductors. But, even if the junction experiment is the strongest method to determine the symmetry of the HTS order parameter, the results have been ambiguous. J. R. Kirtley and C. C. Tsuei thought that the ambiguous results came from the defects inside the HTS, so that they designed an experiment where both clean limit (no defects) and dirty limit (maximal defects) were considered simultaneously. In the experiment, the spontaneous magnetization was clearly observed in YBCO, which supported the *d* symmetry of the order parameter in YBCO. But, since YBCO is orthorhombic, it might inherently have an admixture of *s* symmetry. So, by tuning their technique further, they found that there was an admixture of *s* symmetry in YBCO within about 3%. Also, they found that there was a pure $d_{x^2-y^2}$ order parameter symmetry in the tetragonal $Tl_2Ba_2CuO_6$.

Qualitative Explanation of the Spin-fluctuation Mechanism

Despite all these years, the mechanism of high-T_c superconductivity is still highly controversial, mostly due to the lack of exact theoretical computations on such strongly interacting electron systems. However, most rigorous theoretical calculations, including phenomenological and diagrammatic approaches, converge on magnetic fluctuations as the pairing mechanism for these systems. The qualitative explanation is as follows:

In a superconductor, the flow of electrons cannot be resolved into individual electrons, but instead consists of many pairs of bound electrons, called Cooper pairs. In conventional superconductors, these pairs are formed when an electron moving through the material distorts the surrounding crystal lattice, which in turn attracts another electron and forms a bound pair. This is sometimes called the "water bed" effect. Each Cooper pair requires a certain minimum energy to be displaced, and if the thermal fluctuations in the crystal lattice are smaller than this energy the pair can flow without dissipating energy. This ability of the electrons to flow without resistance leads to super-conductivity.

In a high-T_c superconductor, the mechanism is extremely similar to a conventional superconductor, except, in this case, phonons virtually play no role and their role is replaced by spin-density waves. Just as all known conventional superconductors are strong phonon systems, all known high-T_c superconductors are strong spin-density wave systems, within close vicinity of a magnetic transition to, for example, an antiferromagnet. When an electron moves in a high-T_c superconductor, its spin creates a spin-density wave around it. This spin-density wave in turn causes a near-by electron to fall into the spin depression created by the first electron (water-bed effect again). Hence, again, a Cooper pair is formed. When the system temperature is lowered, more spin density waves and Cooper pairs are created, eventually leading to superconductivity. Note that in high-T_c systems, as these systems are magnetic systems due to the Coulomb interaction, there is a strong Coulomb repulsion between electrons. This Coulomb repulsion prevents pairing of the Cooper pairs on the same lattice site. The pairing of the electrons occur at near-neighbor lattice sites as a result. This is the so-called d-wave pairing, where the pairing state has a node (zero) at the origin.

Examples

Examples of high-T_c cuprate superconductors include $La_{1.85}Ba_{0.15}CuO_4$, and YBCO (Yttrium-Barium-Copper-Oxide), which is famous as the first material to achieve superconductivity above the boiling point of liquid nitrogen.

Transition temperatures of well-known superconductors (Boiling point of liquid nitrogen for comparison)			
Transition temperature (in kelvin)	Transition temperature (in Celsius)	Material	Class
203	−70	H_2S (at 150 GPa pressure)	Hydrogen-based superconductor
195	−78	Sublimation point of dry ice	
184	−89.2	Lowest temperature recorded on Earth	
133	−140	$HgBa_2Ca_2Cu_3O_x$ (HBCCO)	Copper-oxide superconductors
110	−163	$Bi_2Sr_2Ca_2Cu_3O_{10}$ (BSCCO)	
93	−180	$YBa_2Cu_3O_7$ (YBCO)	
90	−183	Boiling point of liquid oxygen	
77	−196	Boiling point of liquid nitrogen	
55	−218	SmFeAs(O,F)	Iron-based superconductors
41	−232	CeFeAs(O,F)	
26	−247	LaFeAs(O,F)	
20	−253	Boiling point of liquid hydrogen	

18	−255	Nb$_3$Sn	Metallic low-temperature superconductors
10	−263	NbTi	
9.2	−263.8	Nb	
4.2	−268.8	Boiling point of liquid helium	
4.2	−268.8	Hg (mercury)	Metallic low-temperature superconductors

Piezoelectricity

Piezoelectricity is the electric charge that accumulates in certain solid materials (such as crystals, certain ceramics, and biological matter such as bone, DNA and various proteins) in response to applied mechanical stress. The word *piezoelectricity* means electricity resulting from pressure. Which means amber, an ancient source of electric charge. Piezoelectricity was discovered in 1880 by French physicists Jacques and Pierre Curie.

The piezoelectric effect is understood as the linear electromechanical interaction between the mechanical and the electrical state in crystalline materials with no inversion symmetry. The piezoelectric effect is a reversible process in that materials exhibiting the direct piezoelectric effect (the internal generation of electrical charge resulting from an applied mechanical force) also exhibit the reverse piezoelectric effect (the internal generation of a mechanical strain resulting from an applied electrical field). For example, lead zirconate titanate crystals will generate measurable piezoelectricity when their static structure is deformed by about 0.1% of the original dimension. Conversely, those same crystals will change about 0.1% of their static dimension when an external electric field is applied to the material. The inverse piezoelectric effect is used in the production of ultrasonic sound waves.

Piezoelectricity is found in useful applications, such as the production and detection of sound, generation of high voltages, electronic frequency generation, microbalances, to drive an ultrasonic nozzle, and ultrafine focusing of optical assemblies. It is also the basis of a number of scientific instrumental techniques with atomic resolution, the scanning probe microscopies, such as STM, AFM, MTA, SNOM, etc., and everyday uses, such as acting as the ignition source for cigarette lighters, and push-start propane barbecues, as well as the time reference source in quartz watches.

History

Discovery and Early Research

The pyroelectric effect, by which a material generates an electric potential in response to a temperature change, was studied by Carl Linnaeus and Franz Aepinus in the mid-18th century. Drawing on this knowledge, both René Just Haüy and Antoine César Becquerel posited a relationship between mechanical stress and electric charge; however, experiments by both proved inconclusive.

The first demonstration of the direct piezoelectric effect was in 1880 by the brothers Pierre Curie and Jacques Curie. They combined their knowledge of pyroelectricity with their understanding of the underlying crystal structures that gave rise to pyroelectricity to predict crystal behavior, and demonstrated the effect using crystals of tourmaline, quartz, topaz, cane sugar, and Rochelle salt (sodium potassium tartrate tetrahydrate). Quartz and Rochelle salt exhibited the most piezoelectricity.

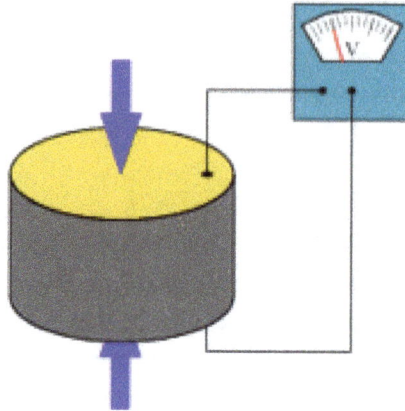

A piezoelectric disk generates a voltage when deformed (change in shape is greatly exaggerated)

The Curies, however, did not predict the converse piezoelectric effect. The converse effect was mathematically deduced from fundamental thermodynamic principles by Gabriel Lippmann in 1881. The Curies immediately confirmed the existence of the converse effect, and went on to obtain quantitative proof of the complete reversibility of electro-elasto-mechanical deformations in piezoelectric crystals.

For the next few decades, piezoelectricity remained something of a laboratory curiosity. More work was done to explore and define the crystal structures that exhibited piezoelectricity. This culminated in 1910 with the publication of Woldemar Voigt's *Lehrbuch der Kristallphysik* (*Textbook on Crystal Physics*), which described the 20 natural crystal classes capable of piezoelectricity, and rigorously defined the piezoelectric constants using tensor analysis.

World War I and Post-war

The first practical application for piezoelectric devices was sonar, first developed during World War I. In France in 1917, Paul Langevin and his coworkers developed an ultrasonic submarine detector. The detector consisted of a transducer, made of thin quartz crystals carefully glued between two steel plates, and a hydrophone to detect the returned echo. By emitting a high-frequency pulse from the transducer, and measuring the amount of time it takes to hear an echo from the sound waves bouncing off an object, one can calculate the distance to that object.

The use of piezoelectricity in sonar, and the success of that project, created intense development interest in piezoelectric devices. Over the next few decades, new piezoelectric materials and new applications for those materials were explored and developed.

Piezoelectric devices found homes in many fields. Ceramic phonograph cartridges simplified player design, were cheap and accurate, and made record players cheaper to maintain and easier to build. The development of the ultrasonic transducer allowed for easy measurement of viscosity and elasticity in fluids and solids, resulting in huge advances in materials research. Ultrasonic time-domain reflectometers (which send an ultrasonic pulse through a material and measure reflections from discontinuities) could find flaws inside cast metal and stone objects, improving structural safety.

World War II and Post-war

During World War II, independent research groups in the United States, Russia, and Japan discovered a new class of synthetic materials, called ferroelectrics, which exhibited piezoelectric constants many times higher than natural materials. This led to intense research to develop barium titanate and later lead zirconate titanate materials with specific properties for particular applications.

One significant example of the use of piezoelectric crystals was developed by Bell Telephone Laboratories. Following World War I, Frederick R. Lack, working in radio telephony in the engineering department, developed the "AT cut" crystal, a crystal that operated through a wide range of temperatures. Lack's crystal didn't need the heavy accessories previous crystal used, facilitating its use on aircraft. This development allowed Allied air forces to engage in coordinated mass attacks through the use of aviation radio.

Development of piezoelectric devices and materials in the United States was kept within the companies doing the development, mostly due to the wartime beginnings of the field, and in the interests of securing profitable patents. New materials were the first to be developed — quartz crystals were the first commercially exploited piezoelectric material, but scientists searched for higher-performance materials. Despite the advances in materials and the maturation of manufacturing processes, the United States market did not grow as quickly as Japan's did. Without many new applications, the growth of the United States' piezoelectric industry suffered.

In contrast, Japanese manufacturers shared their information, quickly overcoming technical and manufacturing challenges and creating new markets. In Japan, a temperature stable crystal cut was developed by Issac Koga. Japanese efforts in materials research created piezoceramic materials competitive to the United States materials but free of expensive patent restrictions. Major Japanese piezoelectric developments included new designs of piezoceramic filters for radios and televisions, piezo buzzers and audio transducers that can connect directly to electronic circuits, and the piezoelectric igniter, which generates sparks for small engine ignition systems (and gas-grill lighters) by compressing a ceramic disc. Ultrasonic transducers that transmit sound waves through air had existed for quite some time but first saw major commercial use in early television remote controls. These transducers now are mounted on several car models as an echolocation device, helping the driver determine the distance from the rear of the car to any objects that may be in its path.

Mechanism

The nature of the piezoelectric effect is closely related to the occurrence of electric dipole moments in solids. The latter may either be induced for ions on crystal lattice sites with asymmetric charge surroundings (as in $BaTiO_3$ and PZTs) or may directly be carried by molecular groups (as in cane sugar). The dipole density or polarization (dimensionality [Cm/m³]) may easily be calculated for crystals by summing up the dipole moments per volume of the crystallographic unit cell. As every dipole is a vector, the dipole density P is a vector field. Dipoles near each other tend to be aligned in regions called Weiss domains. The domains are usually randomly oriented, but can be aligned using the process of *poling* (not the same as magnetic poling), a process by which a strong electric field is applied across the material, usually at elevated temperatures. Not all piezoelectric materials can be poled.

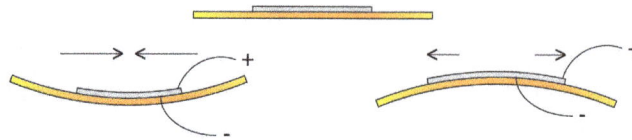

Piezoelectric plate used to convert audio signal to sound waves

Of decisive importance for the piezoelectric effect is the change of polarization P when applying a mechanical stress. This might either be caused by a reconfiguration of the dipole-inducing surrounding or by re-orientation of molecular dipole moments under the influence of the external stress. Piezoelectricity may then manifest in a variation of the polarization strength, its direction or both, with the details depending on: 1. the orientation of P within the crystal; 2. crystal symmetry; and 3. the applied mechanical stress. The change in P appears as a variation of surface charge density upon the crystal faces, i.e. as a variation of the electric field extending between the faces caused by a change in dipole density in the bulk. For example, a 1 cm³ cube of quartz with 2 kN (500 lbf) of correctly applied force can produce a voltage of 12500 V.

Piezoelectric materials also show the opposite effect, called the converse piezoelectric effect, where the application of an electrical field creates mechanical deformation in the crystal.

Mathematical Description

Linear piezoelectricity is the combined effect of

- The linear electrical behavior of the material:

  $$D = \varepsilon E \quad \Rightarrow \quad D_i = \varepsilon_{ij} E_j$$

  where D is the electric charge density displacement (electric displacement), ε is permittivity (free-body dielectric constant), E is electric field strength, and

  $$\nabla \cdot \mathbf{D} = 0, \nabla \times \mathbf{E} = \mathbf{0}.$$

- Hooke's Law for linear elastic materials:

  $$\mathbf{S} = \mathbf{sT} \quad \Rightarrow \quad S_{ij} = s_{ijkl} T_{kl}$$

  where S is strain, s is compliance under short-circuit conditions, T is stress, and

  $$\nabla \cdot \mathbf{T} = \mathbf{0}, \mathbf{S} = \frac{\nabla \mathbf{u} + \mathbf{u} \nabla}{2}.$$

These may be combined into so-called *coupled equations*, of which the strain-charge form is:

$$S = \mathbf{s}T + \partial^t E \quad \Rightarrow \quad S_{ij} = s_{ijkl} T_{kl} + d_{kij} E_k$$
$$D = \partial T + \varepsilon E \quad \Rightarrow \quad D_i = d_{ijk} T_{jk} + \varepsilon_{ij} E_j.$$

In matrix form,

$$\{S\} = \left[s^E\right]\{T\} + [d^t]\{E\}$$

$$\{D\} = [d]\{T\} + \left[\varepsilon^T\right]\{E\},$$

where $[d]$ is the matrix for the direct piezoelectric effect and $[d^t]$ is the matrix for the converse piezoelectric effect. The superscript E indicates a zero, or constant, electric field; the superscript T indicates a zero, or constant, stress field; and the superscript t stands for transposition of a matrix.

Notice that the third order tensor ∂ maps vectors into symmetric matrices. There are no non-trivial rotation-invariant tensors that have this property, which is why there are no isotropic piezoelectric materials.

The strain-charge for a material of the 4mm (C_{4v}) crystal class (such as a poled piezoelectric ceramic such as tetragonal PZT or $BaTiO_3$) as well as the 6mm crystal class may also be written as (ANSI IEEE 176):

$$\begin{bmatrix} S_1 \\ S_2 \\ S_3 \\ S_4 \\ S_5 \\ S_6 \end{bmatrix} = \begin{bmatrix} s_{11}^E & s_{12}^E & s_{13}^E & 0 & 0 & 0 \\ s_{21}^E & s_{22}^E & s_{23}^E & 0 & 0 & 0 \\ s_{31}^E & s_{32}^E & s_{33}^E & 0 & 0 & 0 \\ 0 & 0 & 0 & s_{44}^E & 0 & 0 \\ 0 & 0 & 0 & 0 & s_{55}^E & 0 \\ 0 & 0 & 0 & 0 & 0 & s_{66}^E = 2\left(s_{11}^E - s_{12}^E\right) \end{bmatrix} \begin{bmatrix} T_1 \\ T_2 \\ T_3 \\ T_4 \\ T_5 \\ T_6 \end{bmatrix} + \begin{bmatrix} 0 & 0 & d_{31} \\ 0 & 0 & d_{32} \\ 0 & 0 & d_{33} \\ 0 & d_{24} & 0 \\ d_{15} & 0 & 0 \\ 0 & 0 & 0 \end{bmatrix} \begin{bmatrix} E_1 \\ E_2 \\ E_3 \end{bmatrix}$$

$$\begin{bmatrix} D_1 \\ D_2 \\ D_3 \end{bmatrix} = \begin{bmatrix} 0 & 0 & 0 & 0 & d_{15} & 0 \\ 0 & 0 & 0 & d_{24} & 0 & 0 \\ d_{31} & d_{32} & d_{33} & 0 & 0 & 0 \end{bmatrix} \begin{bmatrix} T_1 \\ T_2 \\ T_3 \\ T_4 \\ T_5 \\ T_6 \end{bmatrix} + \begin{bmatrix} \varepsilon_{11} & 0 & 0 \\ 0 & \varepsilon_{22} & 0 \\ 0 & 0 & \varepsilon_{33} \end{bmatrix} \begin{bmatrix} E_1 \\ E_2 \\ E_3 \end{bmatrix}$$

where the first equation represents the relationship for the converse piezoelectric effect and the latter for the direct piezoelectric effect.

Although the above equations are the most used form in literature, some comments about the notation are necessary. Generally, D and E are vectors, that is, Cartesian tensors of rank 1; and permittivity ε is a Cartesian tensor of rank 2. Strain and stress are, in principle, also rank-2 tensors. But conventionally, because strain and stress are all symmetric tensors, the subscript of strain and stress can be relabeled in the following fashion: $11 \rightarrow 1$; $22 \rightarrow 2$; $33 \rightarrow 3$; $23 \rightarrow 4$; $13 \rightarrow 5$; $12 \rightarrow 6$. (Different conventions may be used by different authors in literature. For example, some use $12 \rightarrow 4$; $23 \rightarrow 5$; $31 \rightarrow 6$ instead.) That is why S and T appear to have the "vector form" of six components. Consequently, s appears to be a 6-by-6 matrix instead of a rank-4 tensor. Such a relabeled notation is often called Voigt notation. Whether the shear strain components S_4, S_5, S_6 are tensor components or engineering strains is another question. In the equation above, they must be engineering strains for the 6,6 coefficient of the compliance matrix to be written as shown, i.e., $2(sE11 - sE12)$. Engineering shear strains are double the value of the corresponding tensor shear, such as $S_6 = 2S_{12}$

and so on. This also means that $s_{66} = 1/G_{12}$, where G_{12} is the shear modulus.

In total, there are four piezoelectric coefficients, d_{ij}, e_{ij}, g_{ij}, and h_{ij} defined as follows:

$$d_{ij} = \left(\frac{\partial D_i}{\partial T_j}\right)^E = \left(\frac{\partial S_j}{\partial E_i}\right)^T$$

$$e_{ij} = \left(\frac{\partial D_i}{\partial S_j}\right)^E = -\left(\frac{\partial T_j}{\partial E_i}\right)^S$$

$$g_{ij} = -\left(\frac{\partial E_i}{\partial T_j}\right)^D = \left(\frac{\partial S_j}{\partial D_i}\right)^T$$

$$h_{ij} = -\left(\frac{\partial E_i}{\partial S_j}\right)^D = -\left(\frac{\partial T_j}{\partial D_i}\right)^S$$

where the first set of four terms corresponds to the direct piezoelectric effect and the second set of four terms corresponds to the converse piezoelectric effect. For those piezoelectric crystals for which the polarization is of the crystal-field induced type, a formalism has been worked out that allows for the calculation of piezoelectrical coefficients d_{ij} from electrostatic lattice constants or higher-order Madelung constants.

Crystal Classes

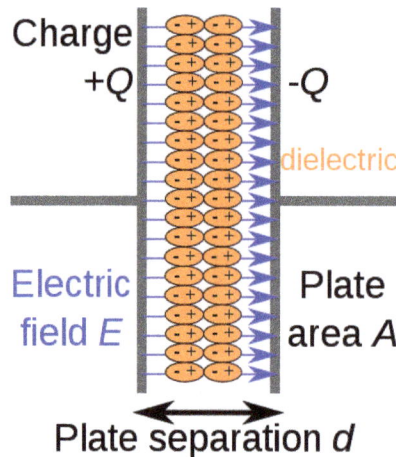

Any spatially separated charge will result in an electric field, and therefore an electric potential. Shown here is a standard dielectric in a capacitor. In a piezoelectric device, mechanical stress, instead of an externally applied voltage, causes the charge separation in the individual atoms of the material.

Of the 32 crystal classes, 21 are non-centrosymmetric (not having a centre of symmetry), and of these, 20 exhibit direct piezoelectricity (the 21st is the cubic class 432). Ten of these represent the polar crystal classes, which show a spontaneous polarization without mechanical stress due to a non-vanishing electric dipole moment associated with their unit cell, and which exhibit pyroelectricity. If the dipole moment can be reversed by the application of an electric field, the material is said to be ferroelectric.

- Polar crystal classes: 1, 2, m, mm2, 4, 4mm, 3, 3m, 6, 6mm.

- Piezoelectric crystal classes: 1, 2, m, 222, mm2, 4, 4, 422, 4mm, 42m, 3, 32, 3m, 6, 6, 622, 6mm, 62m, 23, 43m.

For polar crystals, for which $P \neq 0$ holds without applying a mechanical load, the piezoelectric effect manifests itself by changing the magnitude or the direction of P or both.

For the nonpolar but piezoelectric crystals, on the other hand, a polarization P different from zero is only elicited by applying a mechanical load. For them the stress can be imagined to transform the material from a nonpolar crystal class ($P = 0$) to a polar one, having $P \neq 0$.

Materials

Many materials, both natural and synthetic, exhibit piezoelectricity:

Naturally Occurring Crystals

- Quartz
- Berlinite ($AlPO_4$), a rare phosphate mineral that is structurally identical to quartz
- Sucrose (table sugar)
- Rochelle salt
- Topaz
- Tourmaline-group minerals
- Lead titanate ($PbTiO_3$). Although it occurs in nature as mineral macedonite, it is synthesized for research and applications.

The action of piezoelectricity in Topaz can probably be attributed to ordering of the (F,OH) in its lattice, which is otherwise centrosymmetric: orthorhombic bipyramidal (mmm). Topaz has anomalous optical properties which are attributed to such ordering.

Bone

Dry bone exhibits some piezoelectric properties. Studies of Fukada *et al.* showed that these are not due to the apatite crystals, which are centrosymmetric, thus non-piezoelectric, but due to collagen. Collagen exhibits the polar uniaxial orientation of molecular dipoles in its structure and can be considered as bioelectret, a sort of dielectric material exhibiting quasipermanent space charge and dipolar charge. Potentials are thought to occur when a number of collagen molecules are stressed in the same way displacing significant numbers of the charge carriers from the inside to the surface of the specimen. Piezoelectricity of single individual collagen fibrils was measured using piezoresponse force microscopy, and it was shown that collagen fibrils behave predominantly as shear piezoelectric materials.

The piezoelectric effect is generally thought to act as a biological force sensor. This effect was exploited by research conducted at the University of Pennsylvania in the late 1970s and early 1980s, which established that sustained application of electrical potential could stimulate both resorption and growth (depending on the polarity) of bone in-vivo. Further studies in the 1990s provided the mathematical equation to confirm long bone wave propagation as to that of hexagonal (Class 6) crystals.

Other Natural Materials

Biological materials exhibiting piezoelectric properties include:

- Tendon

- Silk

- Wood due to piezoelectric texture

- Enamel

- Dentin

- DNA

- Viral proteins, including those from bacteriophage. One study has found that thin films of M13 bacteriophage can be used to construct a piezoelectric generator sufficient to operate a liquid crystal display.

Synthetic Crystals

- Langasite ($La_3Ga_5SiO_{14}$), a quartz-analogous crystal

- Gallium orthophosphate ($GaPO_4$), a quartz-analogous crystal

- Lithium niobate ($LiNbO_3$)

- Lithium tantalate ($LiTaO_3$)

Synthetic Ceramics

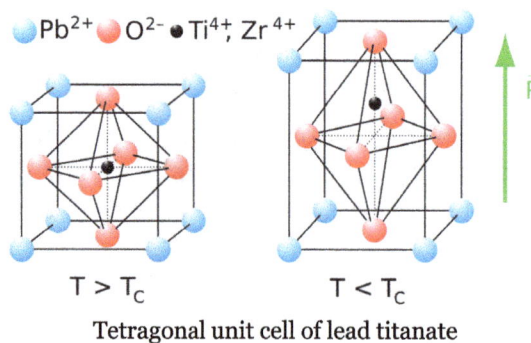

Tetragonal unit cell of lead titanate

Ceramics with randomly oriented grains must be ferroelectric to exhibit piezoelectricity. The macroscopic piezoelectricity is possible in textured polycrystalline non-ferroelectric piezoelectric materials, such as AlN and ZnO. The family of ceramics with perovskite, tungsten-bronze and related structures exhibits piezoelectricity:

- Barium titanate ($BaTiO_3$)—Barium titanate was the first piezoelectric ceramic discovered.

- Lead zirconate titanate ($Pb[Zr_xTi_{1-x}]O_3$ with $0 \leq x \leq 1$)—more commonly known as PZT, lead zirconate titanate is the most common piezoelectric ceramic in use today.

- Potassium niobate ($KNbO_3$)

- Sodium tungstate (Na_2WO_3)

- $Ba_2NaNb_5O_5$

- $Pb_2KNb_5O_{15}$

- Zinc oxide (ZnO)—Wurtzite structure. While single crystals of ZnO are piezoelectric and pyroelectric, polycrystalline (ceramic) ZnO with randomly oriented grains exhibits neither piezoelectric nor pyroelectric effect. Not being ferroelectric, polycrystalline ZnO cannot be poled like barium titanate or PZT. Ceramics and polycrystalline thin films of ZnO may exhibit macroscopic piezoelectricity and pyroelectricity only if they are textured (grains are preferentially oriented), such that the piezoelectric and pyroelectric responses of all individual grains do not cancel. This is readily accomplished in polycrystalline thin films.

Lead-free Piezoceramics

More recently, there is growing concern regarding the toxicity in lead-containing devices driven by the result of restriction of hazardous substances directive regulations. To address this concern, there has been a resurgence in the compositional development of lead-free piezoelectric materials.

- Sodium potassium niobate (($K,Na)NbO_3$). This material is also known as NKN. In 2004, a group of Japanese researchers led by Yasuyoshi Saito discovered a sodium potassium niobate composition with properties close to those of PZT, including a high T_C. Certain compositions of this material have been shown to retain a high mechanical quality factor ($Q_m \approx 900$) with increasing vibration levels, whereas the mechanical quality factor of hard PZT degrades in such conditions. This fact makes NKN a promising replacement for high power resonance applications, such as piezoelectric transformers.

- Bismuth ferrite ($BiFeO_3$) is also a promising candidate for the replacement of lead-based ceramics.

- Sodium niobate $NaNbO_3$

- Bismuth titanate $Bi_4Ti_3O_{12}$

- Sodium bismuth titanate $NaBi(TiO_3)_2$

So far, neither the environmental effect nor the stability of supplying these substances have been measured.

III–V and II–VI Semiconductors

A piezoelectric potential can be created in any bulk or nanostructured semiconductor crystal having non central symmetry, such as the Group III–V and II–VI materials, due to polarization of ions under applied stress and strain. This property is common to both the zincblende and wurtzite crystal structures. To first order, there is only one independent piezoelectric coefficient in zincblende, called e_{14}, coupled to shear components of the strain. In wurtzite, there are instead three independent piezoelectric coefficients: e_{31}, e_{33} and e_{15}. The semiconductors where the strongest piezoelectricity is observed are those commonly found in the wurtzite structure, i.e. GaN, InN, AlN and ZnO. ZnO is the most used material in the recent field of piezotronics.

Since 2006, there have also been a number of reports of strong non linear piezoelectric effects in polar semiconductors. Such effects are generally recognized to be at least important if not of the same order of magnitude as the first order approximation.

Polymers

- Polyvinylidene fluoride (PVDF): PVDF exhibits piezoelectricity several times greater than quartz. Unlike ceramics, where the crystal structure of the material creates the piezoelectric effect, in polymers the intertwined long-chain molecules attract and repel each other when an electric field is applied.

Organic Nanostructures

A strong shear piezoelectric activity was observed in self-assembled diphenylalanine peptide nanotubes (PNTs), indicating electric polarization directed along the tube axis. Comparison with $LiNbO_3$ and lateral signal calibration yields sufficiently high effective piezoelectric coefficient values of at least 60 pm/V (shear response for tubes of ≈200 nm in diameter). PNTs demonstrate linear deformation without irreversible degradation in a broad range of driving voltages.

Application

Currently, industrial and manufacturing is the largest application market for piezoelectric devices, followed by the automotive industry. Strong demand also comes from medical instruments as well as information and telecommunications. The global demand for piezoelectric devices was valued at approximately US$14.8 billion in 2010. The largest material group for piezoelectric devices is piezocrystal, and piezopolymer is experiencing the fastest growth due to its low weight and small size.

Piezoelectric crystals are now used in numerous ways:

High Voltage and Power Sources

Direct piezoelectricity of some substances, like quartz, can generate potential differences of thousands of volts.

- The best-known application is the electric cigarette lighter: pressing the button causes a spring-loaded hammer to hit a piezoelectric crystal, producing a sufficiently high-voltage electric current that flows across a small spark gap, thus heating and igniting the gas. The portable sparkers used to ignite gas stoves work the same way, and many types of gas burners now have built-in piezo-based ignition systems.

- A similar idea is being researched by DARPA in the United States in a project called *Energy Harvesting*, which includes an attempt to power battlefield equipment by piezoelectric generators embedded in soldiers' boots. However, these energy harvesting sources by association affect the body. DARPA's effort to harness 1–2 watts from continuous shoe impact while walking were abandoned due to the impracticality and the discomfort from the additional energy expended by a person wearing the shoes. Other energy harvesting ideas include harvesting the energy from human movements in train stations or other public places and converting a dance floor to generate electricity. Vibrations from industrial machinery can also be harvested by piezoelectric materials to charge batteries for backup supplies or to power low-power microprocessors and wireless radios.

- A piezoelectric transformer is a type of AC voltage multiplier. Unlike a conventional transformer, which uses magnetic coupling between input and output, the piezoelectric transformer uses acoustic coupling. An input voltage is applied across a short length of a bar of piezoceramic material such as PZT, creating an alternating stress in the bar by the inverse piezoelectric effect and causing the whole bar to vibrate. The vibration frequency is chosen to be the resonant frequency of the block, typically in the 100 kilohertz to 1 megahertz range. A higher output voltage is then generated across another section of the bar by the piezoelectric effect. Step-up ratios of more than 1,000:1 have been demonstrated. An extra feature of this transformer is that, by operating it above its resonant frequency, it can be made to appear as an inductive load, which is useful in circuits that require a controlled soft start. These devices can be used in DC–AC inverters to drive cold cathode fluorescent lamps. Piezo transformers are some of the most compact high voltage sources.

Sensors

The principle of operation of a piezoelectric sensor is that a physical dimension, transformed into a force, acts on two opposing faces of the sensing element. Depending on the design of a sensor, different "modes" to load the piezoelectric element can be used: longitudinal, transversal and shear.

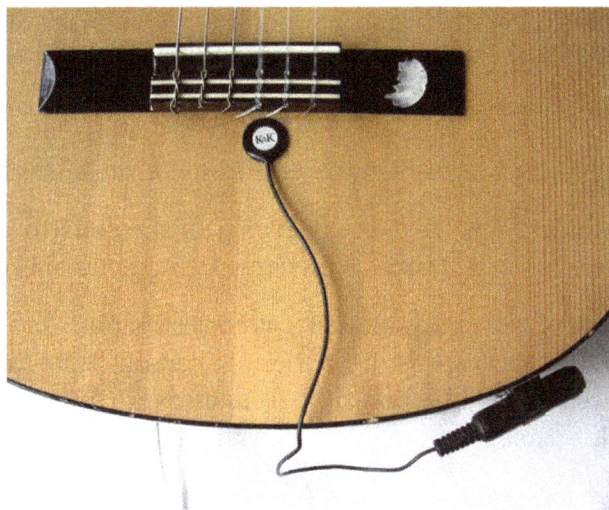

Piezoelectric disk used as a guitar pickup

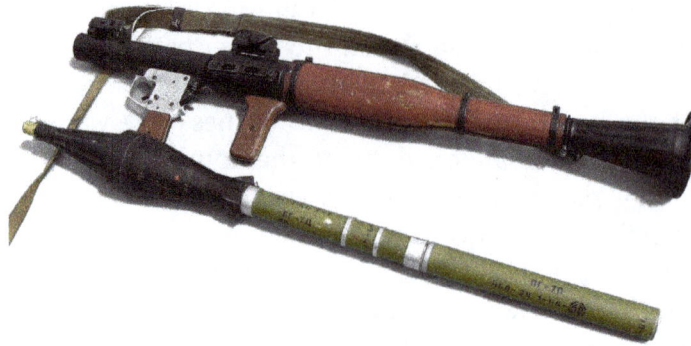

Many rocket-propelled grenades used a piezoelectric fuse. For example: RPG-7.

Detection of pressure variations in the form of sound is the most common sensor application, e.g. piezoelectric microphones (sound waves bend the piezoelectric material, creating a changing voltage) and piezoelectric pickups for acoustic-electric guitars. A piezo sensor attached to the body of an instrument is known as a contact microphone.

Piezoelectric sensors especially are used with high frequency sound in ultrasonic transducers for medical imaging and also industrial nondestructive testing (NDT).

For many sensing techniques, the sensor can act as both a sensor and an actuator – often the term *transducer* is preferred when the device acts in this dual capacity, but most piezo devices have this property of reversibility whether it is used or not. Ultrasonic transducers, for example, can inject ultrasound waves into the body, receive the returned wave, and convert it to an electrical signal (a voltage). Most medical ultrasound transducers are piezoelectric.

In addition to those mentioned above, various sensor applications include:

- Piezoelectric elements are also used in the detection and generation of sonar waves.

- Piezoelectric materials are used in single-axis and dual-axis tilt sensing.

- Power monitoring in high power applications (e.g. medical treatment, sonochemistry and industrial processing).

- Piezoelectric microbalances are used as very sensitive chemical and biological sensors.

- Piezos are sometimes used in strain gauges.

- A piezoelectric transducer was used in the penetrometer instrument on the Huygens Probe.

- Piezoelectric transducers are used in electronic drum pads to detect the impact of the drummer's sticks, and to detect muscle movements in medical acceleromyography.

- Automotive engine management systems use piezoelectric transducers to detect Engine knock (Knock Sensor, KS), also known as detonation, at certain hertz frequencies. A piezoelectric transducer is also used in fuel injection systems to measure manifold absolute pressure (MAP sensor) to determine engine load, and ultimately the fuel injectors milliseconds of on time.

- Ultrasonic piezo sensors are used in the detection of acoustic emissions in acoustic emission testing.

Actuators

Metal disk with piezoelectric disk attached, used in a buzzer.

As very high electric fields correspond to only tiny changes in the width of the crystal, this width can be changed with better-than-μm precision, making piezo crystals the most important tool for positioning objects with extreme accuracy — thus their use in actuators. Multilayer ceramics, using layers thinner than 100 μm, allow reaching high electric fields with voltage lower than 150 V. These ceramics are used within two kinds of actuators: direct piezo actuators and Amplified piezoelectric actuators. While direct actuator's stroke is generally lower than 100 μm, amplified piezo actuators can reach millimeter strokes.

- Loudspeakers: Voltage is converted to mechanical movement of a metallic diaphragm.

- Piezoelectric motors: Piezoelectric elements apply a directional force to an axle, causing it to rotate. Due to the extremely small distances involved, the piezo motor is viewed as a high-precision replacement for the stepper motor.

- Piezoelectric elements can be used in laser mirror alignment, where their ability to move a large mass (the mirror mount) over microscopic distances is exploited to electronically align some laser mirrors. By precisely controlling the distance between mirrors, the laser electronics can accurately maintain optical conditions inside the laser cavity to optimize the beam output.

- A related application is the acousto-optic modulator, a device that scatters light off sound-waves in a crystal, generated by piezoelectric elements. This is useful for fine-tuning a laser's frequency.

- Atomic force microscopes and scanning tunneling microscopes employ converse piezoelectricity to keep the sensing needle close to the specimen.

- Inkjet printers: On many inkjet printers, piezoelectric crystals are used to drive the ejection of ink from the inkjet print head towards the paper.

- Diesel engines: High-performance common rail diesel engines use piezoelectric fuel injectors, first developed by Robert Bosch GmbH, instead of the more common solenoid valve devices.

- Active vibration control using amplified actuators.

- X-ray shutters.

- XY stages for micro scanning used in infrared cameras.

- Moving the patient precisely inside active CT and MRI scanners where the strong radiation or magnetism precludes electric motors.

- Crystal earpieces are sometimes used in old or low power radios.

- High-intensity focused ultrasound for localized heating or creating a localized cavitation can be achieved, for example, in patient's body or in an industrial chemical process.

Frequency Standard

The piezoelectrical properties of quartz are useful as a standard of frequency.

- Quartz clocks employ a crystal oscillator made from a quartz crystal that uses a combination of both direct and converse piezoelectricity to generate a regularly timed series of electrical pulses that is used to mark time. The quartz crystal (like any elastic material) has a precisely defined natural frequency (caused by its shape and size) at which it prefers to oscillate, and this is used to stabilize the frequency of a periodic voltage applied to the crystal.

- The same principle is critical in all radio transmitters and receivers, and in computers where it creates a clock pulse. Both of these usually use a frequency multiplier to reach gigahertz ranges.

Piezoelectric Motors

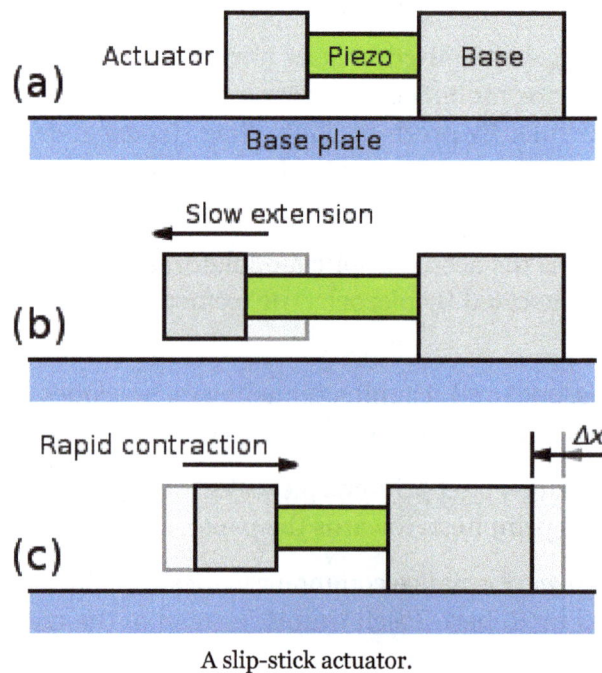

A slip-stick actuator.

Types of piezoelectric motor include:

- The traveling-wave motor used for auto-focus in reflex cameras
- Inchworm motors for linear motion
- Rectangular four-quadrant motors with high power density (2.5 W/cm^3) and speed ranging from 10 nm/s to 800 mm/s.
- Stepping piezo motor, using stick-slip effect.

Aside from the stepping stick-slip motor, all these motors work on the same principle. Driven by dual orthogonal vibration modes with a phase difference of 90°, the contact point between two surfaces vibrates in an elliptical path, producing a frictional force between the surfaces. Usually, one surface is fixed, causing the other to move. In most piezoelectric motors, the piezoelectric crystal is excited by a sine wave signal at the resonant frequency of the motor. Using the resonance effect, a much lower voltage can be used to produce a high vibration amplitude.

A stick-slip motor works using the inertia of a mass and the friction of a clamp. Such motors can be very small. Some are used for camera sensor displacement, thus allowing an anti-shake function.

Reduction of Vibrations and Noise

Different teams of researchers have been investigating ways to reduce vibrations in materials by attaching piezo elements to the material. When the material is bent by a vibration in one direction, the vibration-reduction system responds to the bend and sends electric power to the piezo element to bend in the other direction. Future applications of this technology are expected in cars and houses to reduce noise. Further applications to flexible structures, such as shells and plates, have also been studied for nearly three decades.

In a demonstration at the Material Vision Fair in Frankfurt in November 2005, a team from TU Darmstadt in Germany showed several panels that were hit with a rubber mallet, and the panel with the piezo element immediately stopped swinging.

Piezoelectric ceramic fiber technology is being used as an electronic damping system on some HEAD tennis rackets.

Infertility Treatment

In people with previous total fertilization failure, piezoelectric activation of oocytes together with intracytoplasmic sperm injection (ICSI) seems to improve fertilization outcomes.

Surgery

A recent application of piezoelectric ultrasound sources is piezoelectric surgery, also known as piezosurgery. Piezosurgery is a minimally invasive technique that aims to cut a target tissue with little damage to neighboring tissues. For example, Hoigne *et al.* reported its use in hand surgery for the cutting of bone, using frequencies in the range 25–29 kHz, causing microvibrations of 60–210 μm. It has the ability to cut mineralized tissue without cutting neurovascular tissue and other soft tissue, thereby maintaining a blood-free operating area, better visibility and greater precision.

Potential Applications

In 2015, Cambridge University researchers working in conjunction with researchers from the National Physical Laboratory and Cambridge-based dielectric antenna company Antenova Ltd, using thin films of piezoelectric materials found that at a certain frequency, these materials become not only efficient resonators, but efficient radiators as well, meaning that they can potentially be used as antennas. The researchers found that by subjecting the piezoelectric thin films to an asymmetric excitation, the symmetry of the system is similarly broken, resulting in a corresponding symmetry breaking of the electric field, and the generation of electromagnetic radiation.

In recent years, several attempts at the macro-scale application of the piezoelectric technology have emerged to harvest kinetic energy from walking pedestrians. The piezoelectric floors have been trialed since the beginning of 2007 in two Japanese train stations, Tokyo and Shibuya stations. The electricity generated from the foot traffic is used to provide all the electricity needed to run the automatic ticket gates and electronic display systems. In London, a famous nightclub exploited the piezoelectric technology in its dance floor. Parts of the lighting and sound systems in the club can be powered by the energy harvesting tiles. However, the piezoelectric tile deployed on the ground usually harvests energy from low frequency strikes provided by the foot traffic. This working condition may eventually lead to low power generation efficiency.

In this case, locating high traffic areas is critical for optimization of the energy harvesting efficiency, as well as the orientation of the tile pavement significantly affects the total amount of the harvested energy. A density flow evaluation is recommended to qualitatively evaluate the piezoelectric power harvesting potential of the considered area based on the number of pedestrian crossings per unit time. In X. Li's study, the potential application of a commercial piezoelectric energy harvester in a central hub building at Macquarie University in Sydney, Australia is examined and discussed. Optimization of the piezoelectric tile deployment is presented according to the frequency of pedestrian mobility and a model is developed where 3.1% of the total floor area with the highest pedestrian mobility is paved with piezoelectric tiles. The modelling results indicate that the total annual energy harvesting potential for the proposed optimized tile pavement model is estimated at 1.1 MW h/year, which would be sufficient to meet close to 0.5% of the annual energy needs of the building. In Israel, there is a company which has installed piezoelectric materials under a busy highway. The energy generated is adequate and powers street lights, billboards and signs.

Tyre company Goodyear has plans to develop an electricity generating tyre which has piezoelectric material lined inside it. As the tyre moves, it deforms and thus electricity is generated.

Photovoltaics

The efficiency of a hybrid photovoltaic cell that contains piezoelectric materials can be increased simply by placing it near a source of ambient noise or vibration. The effect was demonstrated with organic cells using zinc oxide nanotubes. The electricity generated by the piezoelectric effect itself is a negligible percentage of the overall output. Sound levels as low as 75 decibels improved efficiency by up to 50%. Efficiency peaked at 10 kHz, the resonant frequency of the nanotubes.

The electrical field set up by the vibrating nanotubes interacts with electrons migrating from the organic polymer layer. This process decreases the likelihood of recombination, in which electrons are energized but settle back into a hole instead of migrating to the electron-accepting ZnO layer.

Compressive Strength

Compressive strength or compression strength is the capacity of a material or structure to withstand loads tending to reduce size, as opposed to tensile strength, which withstands loads tending to elongate. In other words, compressive strength resists compression (being pushed together), whereas tensile strength resists tension (being pulled apart). In the study of strength of materials, tensile strength, compressive strength, and shear strength can be analyzed independently.

Some materials fracture at their compressive strength limit; others deform irreversibly, so a given amount of deformation may be considered as the limit for compressive load. Compressive strength is a key value for design of structures.

Compressive strength is often measured on a universal testing machine; these range from very small table-top systems to ones with over 53 MN capacity. Measurements of compressive strength are affected by the specific test method and conditions of measurement. Compressive strengths are usually reported in relationship to a specific technical standard.

Introduction

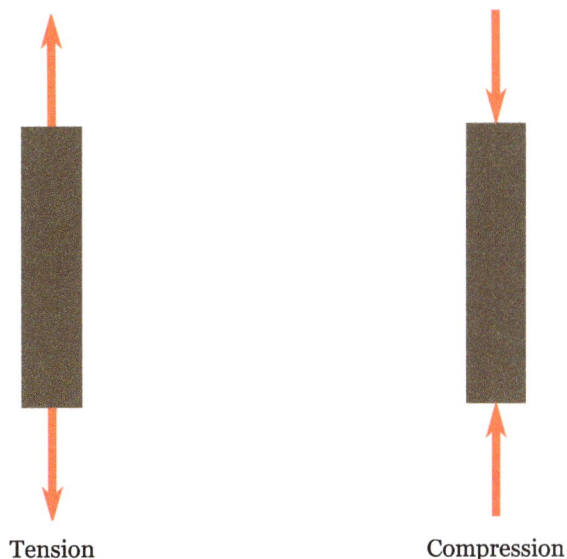

Tension   Compression

When a specimen of material is loaded in such a way that it extends it is said to be in *tension*. On the other hand, if the material compresses and shortens it is said to be in *compression*.

On an atomic level, the molecules or atoms are forced apart when in tension whereas in compression they are forced together. Since atoms in solids always try to find an equilibrium position, and distance between other atoms, forces arise throughout the entire material which oppose both tension or compression. The phenomena prevailing on an atomic level are therefore similar.

The "strain" is the relative change in length under applied stress; positive strain characterises an object under tension load which tends to lengthen it, and a compressive stress that shortens an object gives negative strain. Tension tends to pull small sideways deflections back into alignment, while compression tends to amplify such deflection into buckling.

Compressive strength is measured on materials, components, and structures.

By definition, the ultimate compressive strength of a material is that value of uniaxial compressive stress reached when the material fails completely. The compressive strength is usually obtained experimentally by means of a *compressive test*. The apparatus used for this experiment is the same as that used in a tensile test. However, rather than applying a uniaxial tensile load, a uniaxial compressive load is applied. As can be imagined, the specimen (usually cylindrical) is shortened as well as spread laterally. A stress–strain curve is plotted by the instrument and would look similar to the following:

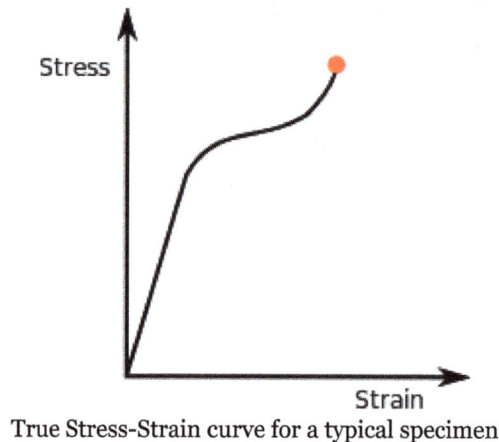

True Stress-Strain curve for a typical specimen

The compressive strength of the material would correspond to the stress at the red point shown on the curve. In a compression test, there is a linear region where the material follows Hooke's Law. Hence for this region $\sigma = E\epsilon$ where this time E refers to the Young's Modulus for compression. In this region, the material deforms elastically and returns to its original length when the stress is removed.

This linear region terminates at what is known as the yield point. Above this point the material behaves plastically and will not return to its original length once the load is removed.

There is a difference between the engineering stress and the true stress. By its basic definition the uniaxial stress is given by:

$$\sigma = \frac{F}{A}$$

where, F = Load applied [N], A = Area [m²]

As stated, the area of the specimen varies on compression. In reality therefore the area is some function of the applied load i.e. A = f(F). Indeed, stress is defined as the force divided by the area at the start of the experiment. This is known as the engineering stress and is defined by,

$$\sigma_e = \frac{F}{A_0}$$

A_0 = Original specimen area [m²]

Correspondingly, the engineering strain would be defined by:

$$\epsilon_e = \frac{l - l_0}{l_0}$$

where l = current specimen length [m] and l_0 = original specimen length [m]

The compressive strength would therefore correspond to the point on the engineering stress strain curve $(\epsilon_e^*, \sigma_e^*)$ defined by

$$\sigma_e^* = \frac{F^*}{A_0}$$

$$\epsilon_e^* = \frac{l^* - l_0}{l_0}$$

where F* = load applied just before crushing and l* = specimen length just before crushing.

Deviation of Engineering Stress from True Stress

Barrelling

In engineering design practice, professionals mostly rely on the engineering stress. In reality, the *true stress* is different from the engineering stress. Hence calculating the compressive strength of a material from the given equations will not yield an accurate result. This is because the cross sectional area A_0 changes and is some function of load A = φ(F).

The difference in values may therefore be summarized as follows:

- On compression, the specimen will shorten. The material will tend to spread in the lateral direction and hence increase the cross sectional area.

- In a compression test the specimen is clamped at the edges. For this reason, a frictional force arises which will oppose the lateral spread. This means that work has to be done to oppose this frictional force hence increasing the energy consumed during the process. This results in a slightly inaccurate value of stress obtained from the experiment.

As a final note, it should be mentioned that the frictional force mentioned in the second point is not constant for the entire cross section of the specimen. It varies from a minimum at the centre, away from the clamps, to a maximum at the edges where it is clamped. Due to this, a phenomenon known as *barrelling* occurs where the specimen attains a barrel shape.

Comparison of Compressive and Tensile Strengths

Concrete and ceramics typically have much higher compressive strengths than tensile strengths. Composite materials, such as glass fiber epoxy matrix composite, tend to have higher tensile strengths than compressive strengths. Metals tend to have tensile and compressive strengths that are very similar.

Typical Values

Material	R_s [MPa]
Porcelain	500
Bone	150
Ice (0 °C)	3
Styrofoam	~1

Compressive Strength of Concrete

Compressive strength is one of the most important engineering property of Concrete which designers are concerned of. It is a standard industrial practice that the concrete is classified based on grades. This grade is nothing but the Compressive Strength of the concrete cube or cylinder. Cube or Cylinder samples are usually tested under a compression testing machine to obtain the compressive strength of concrete. The test requisites differ country to country based on the design code. As per Indian codes, compressive strength of concrete is defined as

The compressive strength of concrete is given in terms of the characteristic compressive strength of 150 mm size cubes tested at 28 days (fck). The characteristic strength is defined as the strength of the concrete below which not more than 5% of the test results are expected to fall.

For design purposes, this compressive strength value is restricted by dividing with a factor of safety, whose value depends on the design philosophy used.

Stiffness

Stiffness is the rigidity of an object — the extent to which it resists deformation in response to an applied force.

The complementary concept is flexibility or pliability: the more flexible an object is, the less stiff it is.

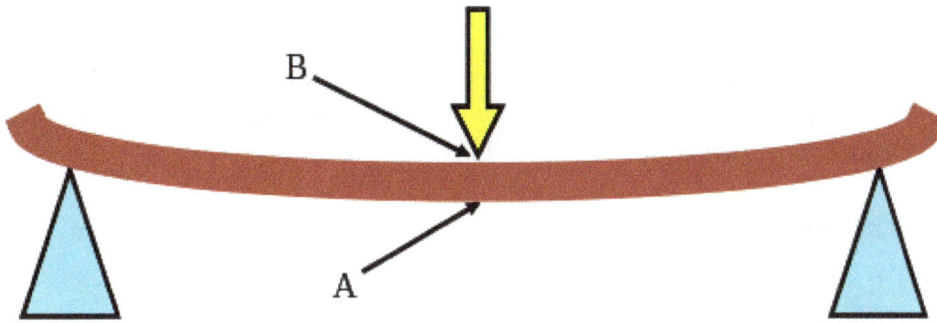

Calculations

The stiffness, k, of a body is a measure of the resistance offered by an elastic body to deformation. For an elastic body with a single degree of freedom (DOF) (for example, stretching or compression of a rod), the stiffness is defined as

$$k = \frac{F}{\delta}$$

where,

> F is the force applied on the body

> δ is the displacement produced by the force along the same degree of freedom (for instance, the change in length of a stretched spring)

In the International System of Units, stiffness is typically measured in newtons per meter. In Imperial units, stiffness is typically measured in pounds(lbs) per inch.

Generally speaking, deflections (or motions) of an infinitesimal element (which is viewed as a point) in an elastic body can occur along multiple DOF (maximum of six DOF at a point). For example, a point on a horizontal beam can undergo both a vertical displacement and a rotation relative to its undeformed axis. When there are M degrees of freedom a M x M matrix must be used to describe the stiffness at the point. The diagonal terms in the matrix are the direct-related stiffnesses (or simply stiffnesses) along the same degree of freedom and the off-diagonal terms are the coupling stiffnesses between two different degrees of freedom (either at the same or different points) or the same degree of freedom at two different points. In industry, the term influence coefficient is sometimes used to refer to the coupling stiffness.

It is noted that for a body with multiple hoj, the equation above generally does not apply since the applied force generates not only the deflection along its own direction (or degree of freedom), but also those along other directions.

For a body with multiple DOF, in order to calculate a particular direct-related stiffness (the diagonal terms), the corresponding DOF is left free while the remaining should be constrained. Under such a condition, the above equation can be used to obtain the direct-related stiffness for the degree of freedom which is unconstrained. The ratios between the reaction forces (or moments) and the produced deflection are the coupling stiffnesses.

A description including all possible stretch and shear parameters is given by the elasticity tensor.

Compliance

The inverse of stiffness is *flexibility* or *compliance* (or sometimes *elastic modulus*), typically measured in units of metres per newton. In rheology it may be defined as the ratio of strain to stress, and so take the units of reciprocal stress, *e.g.* 1/Pa.

Rotational Stiffness

A body may also have a rotational stiffness, k, given by

$$k = \frac{M}{\theta}$$

where

> M is the applied moment
>
> θ is the rotation

In the SI system, rotational stiffness is typically measured in newton-metres per radian.

In the SAE system, rotational stiffness is typically measured in inch-pounds per degree.

Further measures of stiffness are derived on a similar basis, including:

- shear stiffness - ratio of applied shear force to shear deformation

- torsional stiffness - ratio of applied torsion moment to angle of twist

Relationship to Elasticity

The elastic modulus of a material is not the same as the stiffness of a component made from that material. Elastic modulus is a property of the constituent material; stiffness is a property of a structure or component of a structure, and hence it is dependent upon various physical dimensions that describe that component. That is, the modulus is an intensive property of the material; stiffness, on the other hand, is an extensive property of the solid body that is dependent on the material *and* its shape and boundary conditions. For example, for an element in tension or compression, the axial stiffness is

$$k = \frac{AE}{L}$$

where

> A is the cross-sectional area,
>
> E is the (tensile) elastic modulus (or Young's modulus),
>
> L is the length of the element.

Similarly, the rotational stiffness of a straight section is

$$k = \frac{GJ}{L}$$

where

     "J" is the torsion constant for the section,

     "G" is the rigidity modulus of the material

Note that in SI, these units yield $k : \dfrac{N \cdot m}{rad}$. For the special case of unconstrained uniaxial tension or compression, Young's modulus *can* be thought of as a measure of the stiffness of a material.

Applications

The stiffness of a structure is of principal importance in many engineering applications, so the modulus of elasticity is often one of the primary properties considered when selecting a material. A high modulus of elasticity is sought when deflection is undesirable, while a low modulus of elasticity is required when flexibility is needed.

In biology, the stiffness of the extracellular matrix is important for guiding the migration of cells in a phenomenon called durotaxis.

Another application of stiffness finds itself in skin biology. The skin maintains its structure due to its intrinsic tension, contributed to by collagen, an extracellular protein which accounts for approximately 75% of its dry weight. The pliability of skin is a parameter of interest that represents its firmness and extensibility, encompassing characteristics such as elasticity, stiffness, and adherence. These factors are of functional significance to patients. This is of significance to patients with traumatic injuries to the skin, whereby the pliability can be reduced due to the formation and replacement of healthy skin tissue by a pathological scar. This can be evaluated both subjectively, or objectively using a device such as the Cutometer. The Cutometer applies a vacuum to the skin and measures the extent to which it can be vertically distended. These measurements are able to distinguish between healthy skin, normal scarring, and pathological scarring, and the method has been applied within clinical and industrial settings to monitor both pathophysiological sequelae, and the effects of treatments on skin.

Indentation Hardness

Indentation hardness tests are used in mechanical engineering to determine the hardness of a material to deformation. Several such tests exist, wherein the examined material is indented until an impression is formed; these tests can be performed on a macroscopic or microscopic scale.

When testing metals, indentation hardness correlates linearly with tensile strength. This relation permits economically important nondestructive testing of bulk metal deliveries with lightweight, even portable equipment, such as hand-held Rockwell hardness testers.

Material Hardness

As of the direction of materials science continues towards studying the basis of properties on smaller and smaller scales, different techniques are used to quantify material characteristics and tendencies. Measuring mechanical properties for materials on smaller scales, like thin films, can not be done using conventional uniaxial tensile testing. As a result, techniques testing material "hardness" by indenting a material with an impression have been developed to determine such properties.

Hardness measurements quantify the resistance of a material to plastic deformation. Indentation hardness tests compose the majority of processes used to determine material hardness, and can be divided into two classes: *microindentation* and *macroindentation* tests. Microindentation tests typically have forces less than 2 N (0.45 lb$_f$). Hardness, however, cannot be considered to be a fundamental material property. Instead, it represents an arbitrary quantity used to provide a relative idea of material properties. As such, hardness can only offer a comparative idea of the material's resistance to plastic deformation since different hardness techniques have different scales.

The main source of error with indentation tests is the strain hardening effect of the process. However, it has been experimentally determined through "strainless hardness tests" that the effect is minimal with smaller indentations.

Surface finish of the part and the indenter do not have an effect on the hardness measurement, as long as the indentation is large compared to the surface roughness. This proves to be useful when measuring the hardness of practical surfaces. It also is helpful when leaving a shallow indentation, because a finely etched indenter leaves a much easier to read indentation than a smooth indenter.

The indentation that is left after the indenter and load are removed is known to "recover", or spring back slightly. This effect is properly known as *shallowing*. For spherical indenters the indentation is known to stay symmetrical and spherical, but with a larger radius. For very hard materials the radius can be three times as large as the indenter's radius. This effect is attributed to the release of elastic stresses. Because of this effect the diameter and depth of the indentation do contain errors. The error from the change in diameter is known to be only a few percent, with the error for the depth being greater.

Another effect the load has on the indentation is the *piling-up* or *sinking-in* of the surrounding material. If the metal is work hardened it has a tendency to pile up and form a "crater". If the metal is annealed it will sink in around the indentation. Both of these effects add to the error of the hardness measurement.

The equation based definition of hardness is the pressure applied over the contact area between the indenter and the material being tested. As a result hardness values are typically reported in units of pressure, although this is only a "true" pressure if the indenter and surface interface is perfectly flat.

Macroindentation Tests

The term "macroindentation" is applied to tests with a larger test load, such as 1 kgf or more. There are various macroindentation tests, including:

- Vickers hardness test (HV), which has one of the widest scales

- Brinell hardness test (HB)

- Knoop hardness test (HK), for measurement over small areas

- Janka hardness test, for wood

- Meyer hardness test

- Rockwell hardness test (HR), principally used in the USA

- Shore hardness test, for polymers

- Barcol hardness test, for composite materials.

There is, in general, no simple relationship between the results of different hardness tests. Though there are practical conversion tables for hard steels, for example, some materials show qualitatively different behaviors under the various measurement methods. The Vickers and Brinell hardness scales correlate well over a wide range, however, with Brinell only producing overestimated values at high loads.

Microindentation Tests

The term "microhardness" has been widely employed in the literature to describe the hardness testing of materials with low applied loads. A more precise term is "microindentation hardness testing." In microindentation hardness testing, a diamond indenter of specific geometry is impressed into the surface of the test specimen using a known applied force (commonly called a "load" or "test load") of 1 to 1000 gf. Microindentation tests typically have forces of 2 N (roughly 200 gf) and produce indentations of about 50 μm. Due to their specificity, microhardness testing can be used to observe changes in hardness on the microscopic scale. Unfortunately, it is difficult to standardize microhardness measurements; it has been found that the microhardness of almost any material is higher than its macrohardness. Additionally, microhardness values vary with load and work-hardening effects of materials. The two most commonly used microhardness tests are tests that also can be applied with heavier loads as macroindentation tests:

- Vickers hardness test (HV)

- Knoop hardness test (HK)

In microindentation testing, the hardness number is based on measurements made of the indent formed in the surface of the test specimen. The hardness number is based on the applied force divided by the surface area of the indent itself, giving hardness units in kgf/mm². Microindentation hardness testing can be done using Vickers as well as Knoop indenters. For the Vickers test, both the diagonals are measured and the average value is used to compute the Vickers pyramid number. In the Knoop test, only the longer diagonal is measured, and the Knoop hardness is calculated based on the projected area of the indent divided by the applied force, also giving test units in kgf/mm².

The Vickers microindentation test is carried out in a similar manner welling to the Vickers macroindentation tests, using the same pyramid. The Knoop test uses an elongated pyramid to indent material samples. This elongated pyramid creates a shallow impression, which is beneficial for measuring the hardness of brittle materials or thin components. Both the Knoop and Vickers indenters require prepolishing of the surface to achieve accurate results.

Scratch tests at low loads, such as the Bierbaum microcharacter test, performed with either 3 gf or 9 gf loads, preceded the development of microhardness testers using traditional indenters. In

1925, Smith and Sandland of the UK developed an indentation test that employed a square-based pyramidal indenter made from diamond. They chose the pyramidal shape with an angle of 136° between opposite faces in order to obtain hardness numbers that would be as close as possible to Brinell hardness numbers for the specimen. The Vickers test has a great advantage of using one hardness scale to test all materials.The first reference to the Vickers indenter with low loads was made in the annual report of the National Physical Laboratory in 1932. Lips and Sack describes the first Vickers tester using low loads in 1936.

There is some disagreement in the literature regarding the load range applicable to microhardness testing. ASTM Specification E384, for example, states that the load range for microhardness testing is 1 to 1000 gf. For loads of 1 kgf and below, the Vickers hardness (HV) is calculated with an equation, wherein load (L) is in grams force and the mean of two diagonals (d) is in millimeters:

$$HV = 0.0018544 \times \frac{L}{d^2}$$

For any given load, the hardness increases rapidly at low diagonal lengths, with the effect becoming more pronounced as the load decreases. Thus at low loads, small measurement errors will produce large hardness deviations. Thus one should always use the highest possible load in any test. Also, in the vertical portion of the curves, small measurement errors will produce large hardness deviations.

References

- Mourachkine, A. (2004). Room-Temperature Superconductivity. Cambridge International Science Publishing. arXiv:cond-mat/0606187 . ISBN 1-904602-27-4.

- Holler, F. James; Skoog, Douglas A.; Crouch, Stanley R. (2007). "Chapter 1". Principles of Instrumental Analysis (6th ed.). Cengage Learning. p. 9. ISBN 978-0-495-01201-6.

- Gautschi, G. (2002). Piezoelectric Sensorics: Force, Strain, Pressure, Acceleration and Acoustic Emission Sensors, Materials and Amplifiers. Springer. doi:10.1007/978-3-662-04732-3. ISBN 978-3-662-04732-3.

- S. Trolier-McKinstry (2008). "Chapter3: Crystal Chemistry of Piezoelectric Materials". In A. Safari; E.K. Akdoˇgan. Piezoelectric and Acoustic Materials for Transducer Applications. New York: Springer. ISBN 978-0-387-76538-9.

- Becker, Robert O.; Marino, Andrew A. (1982). "Chapter 4: Electrical Properties of Biological Tissue (Piezoelectricity)". Electromagnetism & Life. Albany, New York: State University of New York Press. ISBN 0-87395-560-9.

- Martin Wenham (2001), "Stiffness and flexibility", 200 science investigations for young students, p. 126, ISBN 978-0-7619-6349-3

- Ge, Y. F.; Zhang, F.; Yao, Y. G. (2016). "First-principles demonstration of superconductivity at 280 K in hydrogen sulfide with low phosphorus substitution". Phys. Rev. B. 93 (22): 224513. doi:10.1103/PhysRevB.93.224513.

- Cartlidge, Edwin (18 August 2015). "Superconductivity record sparks wave of follow-up physics". Nature News. Retrieved 18 August 2015.

- Piazza, B. Dalla (15 December 2014). "Fractional excitations in the square-lattice quantum antiferromagnet". Nature Physics 11, 62–68. doi:10.1038/nphys3172. Retrieved 23 December 2014.

- "How electrons split: New evidence of exotic behaviors". Nanowerk. École Polytechnique Fédérale de Lausanne. Dec 23, 2014. Retrieved Dec 23, 2014.

- Jian-Feng Ge; et al. (2014). "Superconductivity in single-layer films of FeSe with a transition temperature above 100 K". 1406: 3435. arXiv:1406.3435 . Bibcode:2014arXiv1406.3435G.

- Bischur, E.; Schwesinger, N. (January 2012). "Organic Piezoelectric Energy Harvesters in Floor". Advanced Materials Research. 433–440: 5848–5853. doi:10.4028/www.scientific.net/AMR.433-440.5848. Retrieved 23 July 2014.

- Urbanek, T; Lee, Johnson. "Column Compression Strength of Tubular Packaging Forms Made of Paper" (PDF). 34,6. Journal of Testing and Evaluation: 31–40. Retrieved 13 May 2014.

- Nedelec, Bernadette; Correa, José; de Oliveira, Ana; LaSalle, Leo; Perrault, Isabelle (2014). "Longitudinal burn scar quantification". Burns. doi:10.1016/j.burns.2014.03.002.

Various Ceramic Materials

Glass-ceramics are produced by controlled crystallizations. Glass-ceramics have the advantage of both glass as well as ceramics. Some of the examples of glass-ceramics are bioactive glass, pyroceram, glass-ceramic-to-metal seals and sitall. This chapter helps the readers in understanding the various materials created by ceramics.

Glass-Ceramic

Glass-ceramics have an amorphous phase and one or more crystalline phases and are produced by a so-called "controlled crystallization" in contrast to a spontaneous crystallization, which is usually not wanted in glass manufacturing. Glass-ceramics have the fabrication advantage of glass, as well as special properties of ceramics. Glass-ceramics do not require brazing but can withstand brazing temperatures up to 700 °C. Glass-ceramics usually have between 30% [m/m] and 90% [m/m] crystallinity and yield an array of materials with interesting properties like zero porosity, high strength, toughness, translucency or opacity, pigmentation, opalescence, low or even negative thermal expansion, high temperature stability, fluorescence, machinability, ferromagnetism, resorbability or high chemical durability, biocompatibility, bioactivity, ion conductivity, superconductivity, isolation capabilities, low dielectric constant and loss, high resistivity and break-down voltage. These properties can be tailored by controlling the base-glass composition and by controlled heat treatment/crystallization of base glass. In manufacturing, glass-ceramics are valued for having the strength of ceramic but the hermetic sealing properties of glass.

Glass-ceramics are mostly produced in two steps: First, a glass is formed by a glass-manufacturing process. The glass is cooled down and is then reheated in a second step. In this heat treatment the glass partly crystallizes. In most cases nucleation agents are added to the base composition of the glass-ceramic. These nucleation agents aid and control the crystallization process. Because there is usually no pressing and sintering, glass-ceramics have, unlike sintered ceramics, no pores.

A wide variety of glass-ceramic systems exists, e.g., the $Li_2O \times Al_2O_3 \times nSiO_2$ system (LAS system), the $MgO \times Al_2O_3 \times nSiO_2$ system (MAS system), the $ZnO \times Al_2O_3 \times nSiO_2$ system (ZAS system).

The LAS system mainly refers to a mix of lithium, silicon, and aluminum oxides with additional components, e.g., glass-phase-forming agents such as Na_2O, K_2O and CaO and refining agents. As nucleation agents most commonly zirconium(IV) oxide in combination with titanium(IV) oxide is used. This important system was studied first and intensively by Hummel, and Smoke.

After crystallization the dominant crystal phase in this type of glass-ceramic is a high-quartz solid solution (HQ s.s.). If the glass-ceramic is subjected to a more intense heat treatment, this HQ s.s. transforms into a keatite-solid solution (K s.s., sometimes wrongly named as beta-spodumene). This transition is non-reversible and reconstructive, which means bonds in the crystal-lattice are broken and new arranged. However, these two crystal phases show a very similar structure as Li could show.

The most interesting properties of these glass-ceramics are their thermomechanical properties. Glass-ceramic from the LAS system is a mechanically strong material and can sustain repeated and quick temperature changes up to 800–1000 °C. The dominant crystalline phase of the LAS glass-ceramics, HQ s.s., has a strong *negative* coefficient of thermal expansion (CTE), keatite-solid solution as still a negative CTE but much higher than HQ s.s. These negative CTEs of the crystalline phase contrasts with the positive CTE of the residual glass. Adjusting the proportion of these phases offers a wide range of possible CTEs in the finished composite. Mostly for today's applications a low or even zero CTE is desired. Also a negative CTE is possible, which means, in contrast to most materials when heated up, such a glass-ceramic contracts. At a certain point, generally between 60% [m/m] and 80% [m/m] crystallinity, the two coefficients balance such that the glass-ceramic as a whole has a thermal expansion coefficient that is very close to zero. Also, when an interface between material will be subject to thermal fatigue, glass-ceramics can be adjusted to match the coefficient of the material they will be bonded to.

Originally developed for use in the mirrors and mirror mounts of astronomical telescopes, LAS glass-ceramics have become known and entered the domestic market through its use in glass-ceramic cooktops, as well as cookware and bakeware or as high-performance reflectors for digital projectors.

Ceramic Matrix Composites

One particularly notable use of glass-ceramics is in the processing of ceramic matrix composites. For many ceramic matrix composites typical sintering temperatures and times cannot be used, as the degradation and corrosion of the constituent fibres becomes more of an issue as temperature and sintering time increase. One example of this is SiC fibres, which can start to degrade via pyrolysis at temperatures above 1470K. One solution to this is to use the glassy form of the ceramic as the sintering feedstock rather than the ceramic, as unlike the ceramic the glass pellets have a softening point and will generally flow at much lower pressures and temperatures. This allows the use of less extreme processing parameters, making the production of many new technologically important fibre-matrix combinations by sintering possible.

Cooktops

Glass-ceramic from the LAS-System is a mechanically strong material and can sustain repeated and quick temperature changes. However, it is not totally unbreakable. Because it is still a brittle material as glass and ceramics are, it can be broken. There have been instances where users reported damage to their cooktops when the surface was struck with a hard or blunt object (such as a can falling from above or other heavy items).

At the same time, it has a very low heat conduction coefficient and can be made nearly transparent (15–20% loss in a typical cooktop) for radiation in the infrared wavelengths.

In the visible range glass-ceramics can be transparent, translucent or opaque and even colored by coloring agents.

Today, there are two major types of electrical stoves with cooktops made of glass-ceramic:

- A glass-ceramic stove uses radiant heating coils or infrared halogen lamps as the heating elements. The surface of the glass-ceramic cooktop above the burner heats up, but the adjacent surface remains cool because of the low heat conduction coefficient of the material.

- An induction stove heats a metal pot's bottom directly through electromagnetic induction.

It is interesting to note that this technology is not entirely new, as glass-ceramic ranges were first introduced in the 1970s using Corningware tops instead of the more durable material used today. These first generation smoothtops were problematic and could only be used with flat-bottomed cookware as the heating was primarily conductive rather than radiative.

Compared to conventional kitchen stoves, glass-ceramic cooktops are relatively simple to clean, due to their flat surface. However, glass-ceramic cooktops can be scratched very easily, so care must be taken not to slide the cooking pans over the surface. If food with a high sugar content (such as jam) spills, it should never be allowed to dry on the surface, otherwise damage will occur.

For best results and maximum heat transfer, all cookware should be flat-bottomed and matched to the same size as the burner zone.

Brands and Manufacturers

Some well-known brands of glass-ceramics are Pyroceram, *Ceran* (cooktops), *Eurokera* (cooktop, stoves and fireplaces), *Zerodur* (telescope mirrors), and *Macor, Kanger* (Glass-Ceramic for cooktop, stoves, microwave and fireplaces door). German manufacturer Schott introduced Zerodur in 1968, Ceran followed in 1971. Nippon Electric Glass of Japan is another worldwide manufacturer of glass ceramics, whose related products in this area include Firelite and Neoceram fire-rated glass. Keralite, manufactured by Vetrotech Saint-Gobain, is a specialty glass-ceramic fire and impact safety rated material for use in fire-rated applications. Glass-ceramics manufactured in the Soviet Union/Russia are known under the name *Sitall*.

The same class of material was also used, until the late 1990s, in CorningWare dishes, which could be taken from the freezer directly to the oven with no risk of thermal shock

Bioactive Glass

Bioactive glasses are a group of surface reactive glass-ceramic biomaterials and include the original bioactive glass, Bioglass. The biocompatibility and bioactivity of these glasses has led them to be investigated extensively for use as implant materials in the human body to repair and replace diseased or damaged bone.

Medical Uses

Bioactive glasses have many applications but these are primarily in the areas of bone repair and bone regeneration via tissue engineering:

- Synthetic bone graft materials for general orthopaedic, craniofacial (bones of the skull and face), maxillofacial and periodontal (the bone structure that supports teeth) repair, chronic osteomyelitis (S53P4 is the only bacterial growth inhibiting bioactive glass). These are available to surgeons in a particulate form

- Cochlear implants

- Bone tissue engineering scaffolds. These are being investigated in many forms, in particular as porous (contains pores into which cells can grow and fluids can travel) 3-dimensional scaffolds

- Treating dentine hypersensitivity and promoting enamel remineralization

Structure

Solid state NMR spectroscopy has been very useful in elucidating the structure of amorphous solids. Bioactive glasses have been studied by ^{29}Si and ^{31}P solid state MAS NMR spectroscopy. The chemical shift from MAS NMR is indicative of the type of chemical species present in the glass. The ^{29}Si MAS NMR spectroscopy showed that Bioglass 45S5 was a Q2 type-structure with a small amount of Q3 ; i.e., silicate chains with a few crosslinks. The ^{31}P MAS NMR revealed predominately Q0 species; i.e., PO_4^{3-}; subsequent MAS NMR spectroscopy measurements have shown that Si-O-P bonds are below detectable levels

Compositions

There have been many variations on the original composition which was Food and Drug Administration (FDA) approved and termed Bioglass. This composition is known as 45S5. Other compositions are in the list below.

- 45S5: 46.1 mol% SiO_2, 26.9 mol% CaO, 24.4 mol% Na_2O and 2.5 mol% P_2O_5. Bioglass

- 58S: 60 mol% SiO_2, 36 mol% CaO and 4 mol% P_2O_5.

- 70S30C: 70 mol% SiO_2, 30 mol% CaO.

- S53P4: 53 mol% SiO_2, 23 mol% Na_2O, 20 mol% CaO and 4 mol% P_2O_5. (S53P4 is the only bacterial growth inhibiting bioactive glass).

Mechanism of Activity

The underlying mechanisms that enable bioactive glasses to act as materials for bone repair have been investigated since the first work of Hench et al. at the University of Florida. Early attention was paid to changes in the bioactive glass surface. Five inorganic reaction stages are commonly thought to occur when a bioactive glass is immersed in a physiological environment:

1) Ion exchange in which modifier cations (mostly Na^+) in the glass exchange with hydronium ions in the external solution.

2) Hydrolysis in which Si-O-Si bridges are broken, forming Si-OH silanol groups, and the glass network is disrupted.

3) Condensation of silanols in which the disrupted glass network changes its morphology to form a gel-like surface layer, depleted in sodium and calcium ions.

4) Precipitation in which an amorphous calcium phosphate layer is deposited on the gel.

5) Mineralization in which the calcium phosphate layer gradually transforms into crystalline hydroxyapatite, that mimics the mineral phase naturally contained with vertebrate bones.

Later, it was discovered that the morphology of the gel surface layer was a key component in determining the bioactive response. This was supported by studies on bioactive glasses derived from sol-gel processing. Such glasses could contain significantly higher concentrations of SiO_2 than traditional melt-derived bioactive glasses and still maintain bioactivity (i.e., the ability to form a mineralized hydroxyapatite layer on the surface). The inherent porosity of the sol-gel-derived material was cited as a possible explanation for why bioactivity was retained, and often enhanced with respect to the melt-derived glass.

Subsequent advances in DNA microarray technology enabled an entirely new perspective on the mechanisms of bioactivity in bioactive glasses. Previously, it was known that a complex interplay existed between bioactive glasses and the molecular biology of the implant host, but the available tools did not provide a sufficient quantity of information to develop a holistic picture. Using DNA microarrays, researchers are now able to identify entire classes of genes that are regulated by the dissolution products of bioactive glasses, resulting in the so-called "genetic theory" of bioactive glasses. The first microarray studies on bioactive glasses demonstrated that genes associated with osteoblast growth and differentiation, maintenance of extracellular matrix, and promotion of cell-cell and cell-matrix adhesion were up-regulated by conditioned cell culture media containing the dissolution products of bioactive glass.

History

Larry Hench and colleagues at the University of Florida first developed these materials in the late 1960s and they have been further developed by his research team at the Imperial College London and other researchers worldwide.

Pyroceram

Pyroceram is the original glass-ceramic material developed and trademarked by Corning Glass in the 1950s.

Development

Its development has been traced to Corning's work in developing photosensitive glass. Corning credits S. Donald Stookey with its discovery; while conducting research in 1953 on a photosensitive lithium silicate glass called Fotoform containing a dispersion of silver nanoparticles, Stookey noted that an accidentally overheated fragment of the glass resisted breakage when dropped. Stookey's initial glass-ceramic became Fotoceram, with $Li_2Si_2O_5$ and quartz as its crystalline phases. Fotoceram evolved into Pyroceram in 1959, with β-spodumene as the crystalline phase, which evolved into the CorningWare line of cookware.

The manufacture of the material involves controlled crystallization. NASA classifies it as a glass-ceramic product.

After about 30 years of informal use as a standard in high heat ($\geq 1000°C$) applications, Pyroceram 9606 was approved by NIST as a certified reference material for thermal conductivity measurements.

Glass-ceramic-to-metal Seals

Glass-to-metal seals have been around for many years, with one of the most common uses being lamp bulb seals. A more recent invention is glass-ceramic-to-metal seals.

Properties

Glass-ceramics are polycrystalline ceramic materials prepared by the controlled crystallization of suitable glasses, normally silicates. Depending on the starting glass composition and the heat-treatment schedule adopted, glass-ceramics can be prepared with tailored thermal expansion characteristics. This makes them ideal for sealing to a variety of different metals, ranging from low expansion Tungsten (W) or Molybdenum (Mo) to high expansion stainless steels and nickel-based superalloys.

Glass-ceramic-to-metal seals offer superior properties over their glass equivalents including more refractory behaviour, in addition to their ability to seal successfully to many different metals and alloys. They have been used in electrical feed-through seals for such applications as vacuum interrupter envelopes and pyrotechnic actuators, in addition to many applications where a higher temperature capability than is possible with glass-to-metal seals is required, including solid oxide fuel cells.

Process

In the formation of a glass-ceramic-to-metal seal, the parts to be joined are first heated, normally under inert atmosphere, in order to melt the glass and allow it to wet and flow into the metal parts, in much the same way as when preparing a more conventional glass-to-metal seal. The temperature is then normally reduced into a temperature regime where many microscopic nuclei are formed in the glass. The temperature is then raised again into a regime where the major crystalline phases can form and grow to create the polycrystalline ceramic material with thermal expansion characteristics matched to that of the particular metal parts.

Examples

The white opaque "glue" between the panel and the funnel of a colour TV cathode ray tube is a devitrified solder glass based on the system $PbO-ZnO-B_2O_3$. While this is a glass-ceramic-to-glass seal, the basic patent of S.A. Claypoole considers glass-ceramic-to-metal seals as well.

Sitall

Sitall aka Sitall CO-115M or Astrositall, is a crystalline glass-ceramic with ultra low coefficient of thermal expansion (CTE). It was originally manufactured in the former Soviet Union and was used in the making of primary mirrors for the Russian Maksutov telescopes, but since dissolution has diminished in quality.

Sitall has a CTE of only $0\pm1.5 \times 10^{-7}$/°C in the temperature range -60 to 60°C, placing it in a rather small group of transparent materials with low CTE such as Pyrex, Zerodur, Cervit and fused quartz.

Materials of low coefficient of thermal expansion are critical in the manufacture of optical elements for telescopes. In segmented mirror telescopes, it is desirable to have this coefficient as near zero as possible, and to have a high degree of homogeneity in the material. The Southern African Large Telescope (SALT) selected Sitall for the manufacture of its 91 primary mirror segments by Lytkarino Optical Glass Factory (LZOS). Using this company was a direct result of increased scientific collaboration between Russia and South Africa since 1994.

Sitall was used for the primary and secondary mirrors of the VLT Survey Telescope.

Sitall has been ballistically tested by The Pentagon for use as a complex composite armour system, intended to resist chemical and kinetic assault.

Glass-to-Metal Seal

Uranium glass used as lead-in seals in a vacuum capacitor

Glass-to-metal seals are a very important element of the construction of vacuum tubes, electric discharge tubes, incandescent light bulbs, glass encapsulated semiconductor diodes, reed switches, pressure tight glass windows in metal cases, and metal or ceramic packages of electronic components.

Properly done, such a seal is hermetic. To achieve such a seal, two properties must hold:

1. The molten glass must be capable of wetting the metal, in order to form a tight bond, and

2. The thermal expansion of the glass and metal must be closely matched so that the seal remains solid as the assembly cools.

When one material goes through a hole in the other, such as a metal wire through a glass bulb, and the inner material's coefficient of thermal expansion is higher than that of the outer, it will shrink more as it cools, cracking the seal. If the inner material's coefficient of expansion is slightly less, the seal will tighten as it cools, which is often beneficial. Since most metals expand much more with heat than most glasses, this is not easy to arrange.

Glass-to-metal Bonds

Glass and metal can bond together by purely mechanical means, which usually gives weaker joints, or by chemical interaction, where the oxide layer on the metal surface forms a strong bond with the glass. The acid-base reactions are main causes of interaction between glass-metal in the presence of metal oxides on the surface of metal. After complete dissolution of the surface oxides into the glass, further progress of interaction depends on the oxygen activity at the interface. The oxygen activity can be increased by diffusion of molecular oxygen through some defects like cracks. Also, reduction of the thermodynamically less stable components in the glass (and releasing the oxygen ions) can increase the oxygen activity at the interface. In other words, the redox reactions are main causes of interaction between glass-metal in the absence of metal oxides on the surface of metal.

For achieving a vacuum-tight seal, the seal must not contain bubbles. The bubbles are most commonly created by gases escaping the metal at high temperature; degassing the metal before its sealing is therefore important, especially for nickel and iron and their alloys. This is achieved by heating the metal in vacuum or sometimes in hydrogen atmosphere or in some cases even in air at temperatures above those used during the sealing process. Oxidizing of the metal surface also reduces gas evolution. Most of the evolved gas is produced due to the presence of carbon impurities in the metals; these can be removed by heating in hydrogen.

The glass-oxide bond is stronger than glass-metal. The oxide forms a layer on the metal surface, with the proportion of oxygen changing from zero in the metal to the stoichiometry of the oxide and the glass itself. A too thick oxide layer tends to be porous on the surface and mechanically weak, flaking, compromising the bond strength and creating possible leakage paths along the metal-oxide interface. Proper thickness of the oxide layer is therefore critical.

Copper

Metallic copper does not bond well to glass. Copper(I) oxide, however, is wetted by molten glass and partially dissolves in it, forming a strong bond. The oxide also bonds well to the underlying metal. But copper(II) oxide causes weak joints that may leak and its formation has to be prevented.

For bonding copper to glass, the surface has to be properly oxidized. The oxide layer has to have the right thickness; too thin oxide would not provide enough material for the glass to anchor to, too thick oxide would fail in the oxide layer, and in both cases the joint would be weak and possibly non-hermetic. To improve the bonding to glass, the oxide layer should be borated; this is achieved by e.g. dipping the hot part into a concentrated solution of borax and then heating it again for certain time. This treatment stabilizes the oxide layer by forming a thin protective layer of sodium borate on its surface, so the oxide does not grow too thick during subsequent handling and joining. The layer should have uniform deep red to purple sheen. The boron oxide from the borated layer diffuses into glass and lowers its melting point. The oxidation occurs by oxygen diffusing through the molten borate layer and forming copper(I) oxide, while formation of copper(II) oxide is inhibited.

The copper-to-glass seal should look brilliant red, almost scarlet; pink, sherry and honey colors are also acceptable. Too thin oxide layer looks light, up to the color of metallic copper. Too thick oxide looks too dark.

Oxygen-free copper has to be used if the metal comes in contact with hydrogen (e.g. in a hydrogen-filled tube or during handling in the flame). Normally, copper contains small inclusions of copper(I) oxide. Hydrogen diffuses through the metal and reacts with the oxide, reducing it to copper and yielding water. The water molecules however can not diffuse through the metal, are trapped in the location of the inclusion, and cause embrittlement.

As copper bonds well to the glass, it is often used for combined glass-metal devices. The ductility of copper can be used for compensation of the thermal expansion mismatch in e.g. the knife-edge seals. For wire feed throughs, dumet wire – nickel-iron alloy plated with copper – is frequently used. Its maximum diameter is however limited to about 0.5 mm due to its thermal expansion.

Copper can be sealed to glass without the oxide layer, but the resulting joint is less strong.

Platinum

Platinum has similar thermal expansion as glass and is well-wetted with molten glass. It however does not form oxides, its bond strength is lower. The seal has metallic color and limited strength.

Gold

Like platinum, gold does not form oxides that could assist in bonding. Glass-gold bonds are therefore metallic in color and weak. Gold tends to be used for glass-metal seals only rarely. Special compositions of soda-lime glasses that match the thermal expansion of gold, containing tungsten trioxide and oxides of lanthanum, aluminum and zirconium, exist.

Silver

Silver forms a thin layer of silver oxide on its surface. This layer dissolves in molten glass and forms silver silicate, facilitating a strong bond.

Nickel

Nickel can bond with glass either as a metal, or via the nickel(II) oxide layer. The metal joint has metallic color and inferior strength. The oxide-layer joint has characteristic green-grey color. Nickel plating can be used in similar way as copper plating, to facilitate better bonding with the underlying metal.

Iron

Iron is only rarely used for feedthroughs, but frequently gets coated with vitreous enamel, where the interface is also a glass-metal bond. The bond strength is also governed by the character of the oxide layer on its surface. A presence of cobalt in the glass leads to a chemical reaction between the metallic iron and cobalt oxide, yielding iron oxide dissolved in glass and cobalt alloying with the iron and forming dendrites, growing into the glass and improving the bond strength.

Iron can not be directly sealed to lead glass, as it reacts with the lead oxide and reduces it to metallic lead. For sealing to lead glasses, it has to be copper-plated or an intermediate lead-free glass has to be used. Iron is prone to creating gas bubbles in glass due to the residual carbon impurities; these can be removed by heating in wet hydrogen. Plating with copper, nickel or chromium is also advised.

Chromium

Chromium is a highly reactive metal present in many iron alloys. Chromium may react with glass, reducing the silicon and forming crystals of chromium silicide growing into the glass and anchoring together the metal and glass, improving the bond strength.

Kovar

Kovar, an iron-nickel-cobalt alloy, has low thermal expansion similar to high-borosilicate glass and is frequently used for glass-metal seals especially for the application in x-ray tubes or glass lasers. It can bond to glass via the intermediate oxide layer of nickel(II) oxide and cobalt(II) oxide; the proportion of iron oxide is low due to its reduction with cobalt. The bond strength is highly dependent on the oxide layer thickness and character. The presence of cobalt makes the oxide layer easier to melt and dissolve in the molten glass. A grey, grey-blue or grey-brown color indicates a good seal. A metallic color indicates lack of oxide, while black color indicates overly oxidized metal, in both cases leading to a weak joint.

Molybdenum

Molybdenum bonds to the glass via the intermediate layer of molybdenum(IV) oxide. Due to its low thermal expansion coefficient, matched to glass, molybdenum, like tungsten, is often used for glass-metal bonds especially in conjunction with aluminium-silicate glass. Its high electrical conductivity makes it superior over nickel-cobalt-iron alloys. It is favored by the lighting industry as feedthroughs for lightbulbs and other devices. Molybdenum oxidizes much faster than tungsten and quickly develops a thick oxide layer that does not adhere well, its oxidation should be therefore limited to just yellowish or at most blue-green color. The oxide is volatile and evaporates as a white smoke above 700 °C; excess oxide can be removed by heating in inert gas (argon) at 1000 °C. Molybdenum strips are used instead of wires where higher currents (and higher cross-sections of the conductors) are needed.

Tungsten

Tungsten bonds to the glass via the intermediate layer of tungsten(VI) oxide. A properly formed bond has characteristic coppery/orange/brown-yellow color in lithium-free glasses; in lithium-containing glasses the bond is blue due to formation of lithium tungstate. Due to its low thermal expansion coefficient, matched to glass, tungsten is frequently used for glass-metal bonds. Tungsten forms satisfying bonds with glasses with similar thermal expansion coefficient such as high-borosilicate glass. The surface of both the metal and glass should be smooth, without scratches. Tungsten has the lowest expansion coefficient of metals and the highest melting point.

Stainless Steel

304 Stainless steel forms bonds with glass via an intermediate layer of chromium(III) oxide and iron(III) oxide. Further reactions of chromium, forming chromium silicide dendrites, are possible. The thermal expansion coefficient of steel is however fairly different from the glass; like with copper, this can be alleviated by using knife-edge (Housekeeper) seals.

Zirconium

Zirconium wire can be sealed to glass with just little treatment – rubbing with abrasive paper and short heating in flame. Zirconium is used in applications demanding chemical resistance or lack of magnetism.

Titanium

Titanium, like zirconium, can be sealed to some glasses with just little treatment.

Indium

Indium and some of its alloys can be used as a solder capable of wetting glass, ceramics, and metals and joining them together. Indium has low melting point and is very soft; the softness allows it to deform plastically and absorb the stresses from thermal expansion mismatches. Due to its very low vapor pressure, indium finds use in glass-metal seals used in vacuum technology.

Gallium

Gallium is a soft metal with melting point at 30 °C. It readily wets glasses and most metals and can be used for seals that can be assembled/disassembled by just slight heating. It can be used as a liquid seal up to high temperatures or even at lower temperatures when alloyed with other metals (e.g. as galinstan).

Mercury

Mercury is a metal liquid at normal temperature. It was used as the earliest glass-to-metal seal and is still in use for liquid seals for e.g. rotary shafts.

Mercury Seal

The first technological use of a glass-to-metal seal was the encapsulation of the vacuum in the barometer by Torricelli. The liquid mercury wets the glass and thus provides for a vacuum tight seal. Liquid mercury was also used to seal the metal leads of early mercury arc lamps into the fused silica bulbs.

A less toxic and more expensive alternative to mercury is gallium.

Mercury and gallium seals can be used for vacuum-sealing rotary shafts.

Platinum Wire Seal

The next step was to use thin platinum wire. Platinum is easily wetted by glass and has a similar coefficient of thermal expansion as typical soda-lime and lead glass. It is also easy to work with because of its non-oxidibility and high melting point. This type of seal was used in scientific equipment throughout the 19th century and also in the early incandescent lamps and radio tubes.

Dumet Wire Seal

In 1911 the Dumet-wire seal was invented which is still the common practice to seal copper leads through soda-lime or lead glass. If copper is properly oxidised before it is wetted by molten glass a vacuum tight seal of good mechanical strength can be obtained. After copper is oxidized, it is often dipped in a borax solution, as borating the copper helps prevents over-oxidation when reintroduced to a flame. Simple copper wire is not usable because its coefficient of thermal expansion is much higher than that of the glass. Thus, on cooling a strong tensile force acts on the glass-to-metal interface and it breaks. Glass and glass-to-metal interfaces are especially sensitive to tensile stress. Dumet-wire is a copper clad wire (about 25% of the weight of the wire is copper) with a core of nickel-iron alloy 42, an alloy with a composition of about 42% nickel. The core has a low coefficient of thermal expansion, allowing for a wire with a coefficient of radial thermal expansion which is slightly lower than the linear coefficient of thermal expansion of the glass, so that the glass-to-metal interface is under a low compression stress. It is not possible to adjust the axial thermal expansion of the wire as well. Because of the much higher mechanical strength of the nickel-iron core compared to the copper, the axial thermal expansion of the Dumet-wire is about the same as of the core. Thus, a shear stress builds up which is limited to a safe value by the low tensile strength of the copper. This is also the reason why Dumet is only useful for wire diameters lower than about 0.5 mm. In a typical Dumet seal through the base of a vacuum tube a short piece of Dumet-wire is butt welded to a nickel wire at one end and a copper wire at the other end. When the base is pressed of lead glass the Dumet-wire and a short part of the nickel and the copper wire are enclosed in the glass. Then the nickel wire and the glass around the Dumet-wire are heated by a gas flame and the glass seals to the Dumet-wire. The nickel and copper do not seal vacuum tight to the glass but are mechanically supported. The butt welding also avoids problems with gas-leakages at the interface between the core wire and the copper.

Copper Tube Seal

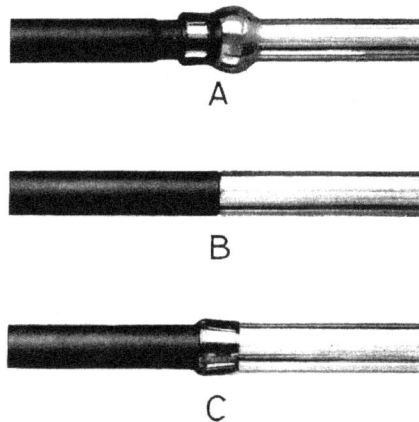

Three types of copper tube seals. In A, the edge of the copper is not in contact with the glass. In B and C, the copper is machined to a sharp knife edge in contact with the glass, with the glass either inside (B) or outside (C) of the copper.

Another possibility to avoid a strong tensile stress when sealing copper through glass is the use of a thin walled copper tube instead of a solid wire. Here a shear stress builds up in the glass-to-metal interface which is limited by the low tensile strength of the copper combined with a low tensile stress. The copper tube is insensitive to high electric current compared to a Dumet-seal because on heating the tensile stress converts into a compression stress which is again limited by the tensile strength of the copper. Also, it is possible to lead an additional solid copper wire through the copper tube. In a later variant, only a short section of the copper tube has a thin wall and the copper tube is hindered to shrink at cooling by a ceramic tube inside the copper tube.

If large parts of copper are to be fitted to glass like the water cooled copper anode of a high power radio transmitter tube or an x-ray tube historically the Houskeeper knife edge seal is used. Here the end of a copper tube is machined to a sharp knife edge, invented by O. Kruh in 1917. In the method described by W.G. Houskeeper the outside or the inside of the copper tube right to the knife edge is wetted with glass and connected to the glass tube. In later descriptions the knife edge is just wetted several millimeters deep with glass, usually deeper on the inside, and then connected to the glass tube.

If copper is sealed to glass, it is an advantage to get a thin bright red Cu_2O containing layer between copper and glass. This is done by borating. After W.J. Scott a copper plated tungsten wire is immersed for about 30 s in chromic acid and then washed thoroughly in running tap water. Then it is dipped into a saturated solution of borax and heated to bright red heat in the oxidizing part of a gas flame. Possibly followed by quenching in water and drying. Another method is to oxidize the copper slightly in a gas flame and then to dip it into borax solution and let it dry. The surface of the borated copper is black when hot and turns to dark wine red on cooling.

It is also possible to make a bright seal between copper and glass where it is possible to see the blank copper surface through the glass, but this gives less adherence than the seal with the red Cu_2O containing layer. If glass is melted on copper in a reducing hydrogen atmosphere the seal is extremely weak. If copper is to be heated in hydrogen-containing atmosphere e.g. a gas flame it needs to be oxygen-free to prevent hydrogen embrittlement. Copper which is meant to be used as an electrical conductor is not necessarily oxygen-free and contains particles of Cu_2O which react with hydrogen that diffuses into the copper to H_2O which cannot diffuse out-off the copper and thus causes embrittlement. The copper usually used in vacuum applications is of the very pure OFHC (oxygen-free-high-conductivity) quality which is both free of Cu_2O and deoxidising additives which might evaporate at high temperature in vacuum.

Copper Disc Seal

In the copper disc seal, as proposed by W.G. Houskeeper, the end of a glass tube is closed by a round copper disc. An additional ring of glass on the opposite side of the disc increases the possible thickness of the disc to more than 0.3 mm. Best mechanical strength is obtained if both sides of the disc are fused to the same type of glass tube and both tubes are under vacuum. The disc seal is of special practical interest because it is a simple method to make a seal to low expansion borosilicate glass without the need of special tools or materials. The keys to success are proper borating, heating of the joint to a temperature as close to the melting point of the copper as possible and to slow down the cooling, at least by packing the assembly into glass wool while it is still red hot.

Matched Seal

In a matched seal the thermal expansion of metal and glass is matched. Copper-plated tungsten wire can be used to seal through borosilicate glass with a low coefficient of thermal expansion which is matched by tungsten. The tungsten is electrolytically copper plated and heated in hydrogen atmosphere to fill cracks in the tungsten and to get a proper surface to easily seal to glass. The borosilicate glass of usual laboratory glassware has a lower coefficient of thermal expansion than tungsten, thus it is necessary to use an intermediate sealing glass to get a stress-free seal.

There are combinations of glass and iron-nickel-cobalt alloys (Kovar) where even the non-linearity of the thermal expansion is matched. These alloys can be directly sealed to glass, but then the oxidation is critical. Also, their low electrical conductivity is a disadvantage. Thus, they are often gold plated. It is also possible to use silver plating, but then an additional gold layer is necessary as an oxygen diffusion barrier to prevent the formation of iron oxide.

While there are Fe-Ni alloys which match the thermal expansion of tungsten at room temperature, they are not useful to seal to glass because of a too strong increase of their thermal expansion at higher temperatures.

Reed switches use a matched seal between an iron-nickel alloy (NiFe 52) and a matched glass. The glass of reed switches is usually green due to its iron content because the sealing of reed switches is done by heating with infrared radiation and this glass shows a high absorption in the near infrared.

The electrical connections of high-pressure sodium vapour lamps, the yellow lamps for street lighting, are made of niobium alloyed with 1% of zirconium.

Historically, some television cathode ray tubes were made by using ferric steel for the funnel and glass matched in expansion to ferric steel. The steel plate used had a diffusion layer enriched with chromium at the surface made by heating the steel together with chromium oxide in a HCl-containing atmosphere. In contrast to copper, pure iron does not bond strongly to silicate glass. Also, technical iron contains some carbon which forms bubbles of CO when it is sealed to glass under oxidizing conditions. Both are a major source of problems for the technical enamel coating of steel and make direct seals between iron and glass unsuitable for high vacuum applications. The oxide layer formed on chromium-containing steel can seal vacuum tight to glass and the chromium strongly reacts with carbon. Silver-plated iron was used in early microwave tubes.

It is possible to make matched seals between copper or austenitic steel and glass, but silicate glass with that high thermal expansion is especially fragile and has a low chemical durability.

Molybdenum Foil Seal

Another widely used method to seal through glass with low coefficient of thermal expansion is the use of strips of thin molybdenum foil. This can be done with matched coefficients of thermal expansion or unmatched after W.G. Houskeeper. Then the edges of the strip also have to be knife sharp. The disadvantage here is that the tip of the edge which is a local point of high tensile stress reaches through the wall of the glass container. This can lead to low gas leakages. In the tube to tube knife edge seal the edge is either outside, inside, or buried into the glass wall.

Compression Seal

Another possibility of seal construction is the compression seal. This type of glass-to-metal seal can be used to feed through the wall of a metal container. Here the wire is usually matched to the glass which is inside of the bore of a strong metal part with higher coefficient of thermal expansion. Compression seals can withstand extremely high pressures and physical stress such as mechanical and thermal shock. Because glass is extremely strong in compression, compression seals can withstand very high pressures.

Silver Chloride

Silver chloride, which melts at 457 C bonds to glass, metals and other materials and has been used for vacuum seals. Even if it can be a convenient way to seal metal into glass it will not be a true glass to metal seal but rather a combination of a glass to silver chloride and a silver chloride to glass bond, an inorganic alternative to wax or glue bonds.

Design Aspects

Also the mechanical design of a glass-to-metal seal has an important influence on the reliability of the seal. In practical glass-to-metal seals cracks usually start at the edge of the interface between glass and metal either inside or outside the glass container. If the metal and the surrounding glass are symmetric the crack propagates in an angle away from the axis. So, if the glass envelope of the metal wire extends far enough from the wall of the container the crack will not go through the wall of the container but it will reach the surface on the same side where it started and the seal will not leak despite the crack.

Another important aspect is the wetting of the metal by the glass. If the thermal expansion of the metal is higher than the thermal expansion of the glass like with the Housekeeper seal, a high contact angle (bad wetting) means that there is a high tensile stress in the surface of the glass near the metal. Such seals usually break inside the glass and leave a thin cover of glass on the metal. If the contact angle is low (good wetting) the surface of the glass is everywhere under compression stress like an enamel coating. Ordinary soda-lime glass does not flow on copper at temperatures below the melting point of the copper and, thus, does not give a low contact angle. The solution is to cover the copper with a solder glass which has a low melting point and does flow on copper and then to press the soft soda-lime glass onto the copper. The solder glass must have a coefficient of thermal expansion which is equal or a little lower than that of the soda-lime glass. Classically high lead containing glasses are used, but it is also possible to substitute these by multi-component glasses e.g. based on the system Li_2O-Na_2O-K_2O-CaO-SiO_2-B_2O_3-ZnO-TiO_2-BaO-Al_2O_3.

Terracotta

Terracotta, terra cotta or terra-cotta (pronounced Italian: "baked earth", from the Latin *terra cocta*), a type of earthenware, is a clay-based unglazed or glazed ceramic, where the fired body is porous. Terracotta is the term normally used for sculpture made in earthenware, and also for various utilitarian uses including vessels (notably flower pots), water and waste water pipes, roofing tiles, bricks, and surface embellishment in building construction. The term is also used to refer to the natural, brownish orange color, of most terracotta, which varies considerably.

Terracotta head from Akhnoor, Kashmir. Head dates back to 6th century AD. On display in Prince of Wales Museum

International Gothic bust of the Virgin Mary, Bohemia, c. 1390–95, terracotta with polychromy

Sculpture of Hanuman in unglazed terracotta

Terracotta Army in Xi'an, China

Roundel by Luca and Andrea della Robbia

This article covers the senses of terracotta as a medium in sculpture, as in the Terracotta Army and Greek terracotta figurines, and architectural decoration. Asian and European sculpture in porcelain is not covered. Glazed architectural terracotta and its unglazed version as exterior surfaces for buildings were used in Asia for some centuries before becoming popular in the West in the 19th century. Architectural terracotta can also refer to decorated ceramic elements such as antefixes and revetments, which made a large contribution to the appearance of temples and other buildings in the classical architecture of Europe, as well as in the Ancient Near East.

In archaeology and art history, "terracotta" is often used to describe objects such as figurines not made on a potter's wheel. Vessels and other objects that are or might be made on a wheel from the same material are called earthenware pottery; the choice of term depends on the type of object rather than the material or firing technique. Unglazed pieces, and those made for building construction and industry, are also more likely to be referred to as terracotta, whereas tableware and other vessels are called earthenware (though sometimes terracotta if unglazed), or by a more precise term such as faience.

Production and Properties

An appropriate refined clay is formed to the desired shape. After drying it is placed in a kiln or atop combustible material in a pit, and then fired. The typical firing temperature is around 1,000 °C (1,830 °F), though it may be as low as 600 °C (1,112 °F) in historic and archaeological examples. The iron content, reacting with oxygen during firing, gives the fired body a reddish color, though the overall color varies widely across shades of yellow, orange, buff, red, "terracotta", pink, grey or brown. In some contexts, such as Roman figurines, white-colored terracotta is known as pipeclay, as such clays were later preferred for tobacco pipes, normally made of clay until the 19th century.

Fired terracotta is not watertight, but surface-burnishing the body before firing can decrease its porousness and a layer of glaze can make it watertight. It is suitable for use below ground to carry pressurized water (an archaic use), for garden pots or building decoration in many environments, and for oil containers, oil lamps, or ovens. Most other uses, such as for tableware, sanitary piping, or building decoration in freezing environments, require the material to be glazed. Terracotta, if uncracked, will ring if lightly struck.

Painted ("polychrome") terracotta is typically first covered with a thin coat of gesso, then painted. It has been very widely used but the paint is only suitable for indoor positions and is much less durable than fired colors in or under a ceramic glaze. Terracotta sculpture was very rarely left in its "raw" fired state in the West until the 18th century.

History

Terracotta/earthenware was the only known type of ceramic produced by Western and pre-Columbian people until the 14th century, when imported European fired stoneware began production. Terracotta has been used throughout history for sculpture and pottery as well as for bricks and roof shingles. In ancient times, the first clay sculptures were dried (baked) in the sun after being formed. They were later placed in the ashes of open hearths to harden, and finally kilns were used, similar to those used for pottery today. However, only after firing to high temperature would it be classed as a ceramic material.

In Art History

Terracotta female figurines were uncovered by archaeologists in excavations of Mohenjo-daro, Pakistan (3000–1500 BC). Along with phallus-shaped stones, these suggest some sort of fertility cult and a belief in a mother goddess. The Burney Relief is an outstanding terracotta plaque from Ancient Mesopotamia of about 1950 BC. In Mesoamerica, the great majority of Olmec figurines were in terracotta. Many ushabti mortuary statuettes were also made of terracotta in Ancient Egypt.

The Ancient Greeks' Tanagra figurines were mass-produced mold-cast and fired terracotta figurines, that seem to have been widely affordable in the Hellenistic period, and often purely decorative in function. They were part of a wide range of Greek terracotta figurines, which included larger and higher-quality works such as the Aphrodite Heyl; the Romans too made great numbers of small figurines, often religious. Etruscan art often used terracotta in preference to stone even for larger statues, such as the near life-size Apollo of Veii and the *Sarcophagus of the Spouses*. Campana reliefs are Ancient Roman terracotta reliefs, originally mostly used to make friezes for the outside of buildings, as a cheaper substitute for stone.

Indian sculpture made heavy use of terracotta from as early as the Indus Valley Civilization (with stone and metal sculpture being rather rare), and in more sophisticated areas had largely abandoned modeling for using molds by the 1st century BC. This allows relatively large figures, nearly up to life-size, to be made, especially in the Gupta period and the centuries immediately following it. Several vigorous local popular traditions of terracotta folk sculpture remain active today, such as the Bankura horses.

Precolonial West African sculpture also made extensive use of terracotta. The regions most recognized for producing terracotta art in that part of the world include the Nok culture of central and north-central Nigeria, the Ife/Benin cultural axis in western and southern Nigeria (also noted for its exceptionally naturalistic sculpture), and the Igbo culture area of eastern Nigeria, which excelled in terracotta pottery. These related, but separate, traditions also gave birth to elaborate schools of bronze and brass sculpture in the area.

Chinese sculpture made great use of terracotta, with and without glazing and colour, from a very early date. The famous Terracotta Army of Emperor Qin Shi Huang, 209–210 BC, was somewhat untypical, and two thousand years ago reliefs were more common, in tombs and elsewhere. Later

Buddhist figures were often made in painted and glazed terracotta, with the Yixian glazed pottery luohans, probably of 1150–1250, now in various Western museums, among the finest examples. Brick-built tombs from the Han dynasty were often finished on the interior wall with bricks decorated on one face; the techniques included molded reliefs. Later tombs contained many figures of protective spirits and animals and servants for the afterlife, including the famous horses of the T'ang dynasty; as an arbitrary matter of terminology these tend not to be referred to as terracottas.

European medieval art made little use of terracotta sculpture, until the late 14th century, when it became used in advanced International Gothic workshops in parts of Germany. The Virgin illustrated at the start of the article from Bohemia is the unique example known from there. A few decades later there was a revival in the Italian Renaissance, inspired by excavated classical terracottas as well as the German examples, which gradually spread to the rest of Europe. In Florence Luca della Robbia (1399/1400–1482) was a sculptor who founded a family dynasty specializing in glazed and painted terracotta, especially large roundels which were used to decorate the exterior of churches and other buildings. These used the same techniques as contemporary maiolica and other tin-glazed pottery. Other sculptors included Pietro Torrigiano (1472–1528), who produced statues, and in England busts of the Tudor royal family. The unglazed busts of the Roman Emperors adorning Hampton Court Palace, by Giovanni da Maiano, 1521, were another example of Italian work in England. They were originally painted but this has now been lost from weathering.

Clodion, *The River Rhine Separating the Waters*, 1765, 27.9×45.7×30.5 cm (11×18×12 in)

In the 18th-century unglazed terracotta, which had long been used for preliminary clay models or maquettes that were then fired, became fashionable as a material for small sculptures including portrait busts. It was much easier to work than carved materials, and allowed a more spontaneous approach by the artist. Claude Michel (1738–1814), known as Clodion, was an influential pioneer in France. John Michael Rysbrack (1694–1770), a Flemish portrait sculptor working in England, sold his terracotta *modelli* for larger works in stone, and produced busts only in terracotta. In the next century the French sculptor Albert-Ernest Carrier-Belleuse made many terracotta pieces, but possibly the most famous is The Abduction of Hippodameia depicting the Greek mythological scene of a centaur kidnapping Hippodameia on her wedding day.

Architecture

Terracotta tiles have a long history in many parts of the world, covered in that article. Many ancient and traditional roofing styles included more elaborate sculptural elements than the plain roof tiles, such as Chinese Imperial roof decoration and the antefix of western classical architecture. In India West Bengal made a speciality of terracotta temples, with the sculpted decoration from the same material as the main brick construction.

Imperial roof decoration in the Forbidden City

In the 19th century the possibilities of terracotta decoration of buildings were again appreciated by architects, often using thicker pieces of terracotta, and surfaces that are not flat. The American architect Louis Sullivan is well known for his elaborate glazed terracotta ornamentation, designs that would have been impossible to execute in any other medium. Terracotta and tile were used extensively in the town buildings of Victorian Birmingham, England. By about 1930 the widespread use of concrete and Modernist architecture largely ended the use of terracotta in architecture.

Advantages in Sculpture

As compared to bronze sculpture, terracotta uses a far simpler and quicker process for creating the finished work with much lower material costs. The easier task of modelling, typically with a limited range of knives and wooden shaping tools, but mainly using the fingers, allows the artist to take a more free and flexible approach. Small details that might be impractical to carve in stone, of hair or costume for example, can easily be accomplished in terracotta, and drapery can sometimes be made up of thin sheets of clay that make it much easier to achieve a realistic effect.

Reusable mold-making techniques may be used for production of many identical pieces. Compared to marble sculpture and other stonework the finished product is far lighter and may be further painted and glazed to produce objects with color or durable simulations of metal patina. Robust durable works for outdoor use require greater thickness and so will be heavier, with more care needed in the drying of the unfinished piece to prevent cracking as the material shrinks. Structural considerations are similar to those required for stone sculpture; there is a limit on the stress that can be imposed on terracotta, and terracotta statues of unsupported standing figures are limited to well under life-size unless extra structural support is added. This is also because large figures are extremely difficult to fire, and surviving examples often show sagging or cracks. The Yixian figures were fired in several pieces, and have iron rods inside to hold the structure together.

Rare terracotta image of Isis lamenting the loss of Osiris (Eighteenth Dynasty, Egypt) Musée du Louvre, Paris

Wealthy 'Middle-class' women: so-called Tanagra figurine, ancient Greece, 325-150 BC, Altes Museum

Indian terracotta figures, Gupta dynasty

Maximilien Robespierre, unglazed bust by Claude-André Deseine, 1791

The Etruscan "Sarcophagus of the Spouses", at the National Etruscan Museum, c 520 BC

Han dynasty "tomb brick" relief

The Natural History Museum in London has an ornate terracotta façade typical of high Victorian architecture. The carvings represent the contents of the Museum.

Terracotta temple, Bishnupur, India, a famous centre for terracotta temples

Hindu temple, 1739, Kalna, India

he Bell Edison Telephone Building, Birmingham, England.

Architectural Terracotta

Architectural terracotta refers either to decorated ceramic elements such as antefixes and revetments, usually brightly painted, which made a large contribution to the appearance of temples and other buildings in the classical architecture of Europe, as well as in the Ancient Near East, or to forms of ceramic skins used to surface buildings from the 19th century onwards. In the latter sense unglazed architectural terracotta became fashionable as an architectural ceramic construction material in England in the 1860s, and in the United States in the 1870s. Glazed architectural terracotta was developed a little later. Both were generally used as a decorative skin to cover or supplement brick and tiles of similar colour in late Victorian buildings.

St Stephen and All Martyrs' Church, Lever Bridge, Bolton, England

The architectural detail on the facade of the building represents the contents of the Museum

Terracotta had been used architecturally before this in Germany from 1824 by Karl Friedrich Schinkel. Edmund Sharpe designed and oversaw the construction of the first church built almost exclusively of the material, St Stephen and All Martyrs' Church, Lever Bridge in Bolton, erected 1842–45. Henry Cole, secretary to the Science and Arts Department of the UK adopted terracotta for the building which is now the Victoria and Albert Museum (1859–71) and then the Royal Albert Hall (1867–71), both in London. Alfred Waterhouse used it in his designs when in business in Manchester from 1853 and London from 1865. He used a combination of buff and blue-grey terracotta in his Natural History Museum in London. The colour of terracotta varies with the source of the clay. In Britain, London clay gives a pale pink or buff colour, whereas the Ruabon (North Wales) clay gives a bright red.

Terracotta had the advantage of being cheap and light. It was adaptable to mass-production techniques for stock shapes, although the plaster moulds had a limited capability for re-use. Additionally it could be freely worked by craftsmen to make custom-sculptured adornments and plaques. It was accepted as a material by the Arts and Crafts movement because despite seeming a mass-produced material it was handmade and designed by craftsmen. It had a manufacture time of about eight weeks and each piece had to be made over-size to allow for shrinkage as the clay body dried. To avoid cracking the pieces had to be quite thin. They were filled with concrete as they were applied to buildings.

The disadvantage of terracotta, apart from its rather uniform colour in a given district, was that it was not easy to keep clean. Town smoke made it blacken. A more modern phenomenon is the growth of naturally seeded plants and small trees which grow in the nooks and crannies of the intricate designs high above the streets now that the Victorian pollution has gone. Unglazed terracotta went out of fashion from around the 1890s, giving way to glazed architectural terra-cotta, or *faience* as it is known in Britain, which does not attract grime and is easy to clean, giving way to a more colourful architecture.

Glazed tiles had been widely used as a surfacing material before, especially in Islamic architecture, but architectural terracotta typically used relief or three-dimensional decoration in a way not normal with tiling, to achieve an effect similar to carved stone but with less effort and expense.

Manufacturers

- John Marriott Blashfield
- Burmantofts Pottery
- Fambrini & Daniels, Lincoln
- Gibbs and Canning Limited
- Gladding, McBean, California, USA
- Perth Amboy Terra Cotta Company of Perth Amboy, New Jersey, USA
- Boston Valley Terra Cotta, Buffalo, NY, USA

Ceramic Foam

Ceramic foam is a tough foam made from ceramics. Manufacturing techniques include impregnating open-cell polymer foams internally with ceramic slurry and then firing in a kiln, leaving only ceramic material. The foams may consist of several ceramic materials such as aluminium oxide, a common high-temperature ceramic, and gets insulating properties from the many tiny air-filled voids within the material.

The foam can be used not only for thermal insulation, but for a variety of other applications such as acoustic insulation, absorption of environmental pollutants, filtration of molten metal alloys, and as substrate for catalysts requiring large internal surface area.

It has been used as stiff lightweight structural material, specifically for support of reflecting telescope mirrors.

Properties

Ceramic foams are hardened ceramics with pockets of air or another gas trapped in pores throughout the body of the material. These materials can be fabricated as high as 94 to 96% air by volume with temperature resistances as high as 1700 °C. Because many ceramics are already oxides or other inert compounds, there is no danger of oxidation or reduction of the material.

Previously, pores had been avoided in ceramic components due to their brittle properties. However, in practice ceramic foams have somewhat advantageous mechanical properties compared to bulk ceramics. One example is crack propagation, given by:

$$\sigma_t = 2\sigma \left(\frac{a}{r} \right)^{\frac{1}{2}}$$

where σ_t is the stress at the tip of the crack, σ is the applied stress, a is the crack size and r is the radius of curvature. For certain stress applications, this means ceramic foams actually outperform bulk ceramics because the porous pockets of air act to blunt the crack tip radius, leading to a disruption of its propagation and a decrease in the likelihood of failure.

Manufacturing

Much like metal foams, there are a number of accepted methods for creating ceramic foams. One of the earliest and still most common is the polymeric sponge method. A polymeric sponge is covered with a ceramic in suspension, and after rolling to ensure all pores have been filled, the ceramic-coated sponge is dried and pyrolysed to decompose the polymer, leaving only the porous ceramic structure. The foam must then be sintered for final densification. This method is widely used because it is effective with any ceramic able to be suspended; however, large amounts of gaseous byproducts are released and cracking due to differences in thermal expansion coefficients are common.

While the above are both based on the use of a sacrificial template, there are also direct foaming methods that can be used. These methods involve pumping air into a suspended ceramic before setting and sintering. This is difficult because wet foams are thermodynamically unstable and can end up with very large pores after setting.

A recent method of creating aluminum oxide foams has also been developed. This technique involves heating crystals with the metal and forming compounds until a solution is created. At this point, polymer chains form and grow, causing the entire mixture to separate into a solvent and polymer. As the mixture begins to boil, air bubbles are trapped in solution and locked in to place as the material is heated and polymer is burned off.

Use

Insulation

Due to ceramics' extremely low thermal conductivity, the most obvious use of a ceramic is as an insulation material. Ceramic foams are notable in this regard because their composition by very common compounds, such as aluminum oxide, makes them completely harmless, unlike asbestos and other ceramic fibers. Their high strength and hardness also allows them to be used as structural materials for low stress applications.

Electronics

With easily controlled porosities and microstructures, ceramic foams have seem growing use in evolving electronics applications. These applications include electrodes, and scaffolds for solid oxide fuel cells and batteries. Foams can also be used as cooling components for electronics by separating a pumped coolant from the circuits themselves. For this application, silica, aluminum oxide, and aluminum borosilicate fibers can be used.

Pollution Control

Ceramic foams have been proposed as a means of pollutant control, particularly for particulate matter from engines. They are effective because the voids can capture particulates as well as support a catalyst that can induce oxidation of the captured particulates. Due to the easy means of deposition of other materials within ceramic foams, these oxidation-inducing catalysts can easily be distributed through the entire foam, increasing effectiveness.

References

- Merrill L. Minges; Handbook Committee (1989). Electronic Materials Handbook: Packaging. CRC Press. ISBN 978-0-87170-285-2.

- Fred Rosebury (1992-12-31). Handbook of electron tube and vacuum techniques. American Institute . of Physics. ISBN 978-1-56396-121-2.

- Kohl, Walter Heinrich (1967). Handbook of materials and techniques for vacuum devices. American Institute of Physics. ISBN 978-1-56396-387-2.

- Schultz, Ellen (ed). Gothic and Renaissance Art in Nuremberg, 1986, New York, Metropolitan Museum of Art, ISBN 9780870994661, google books

- M. Montazerian, S.P. Singh & E.D. Zanotto, "An Analysis of Glass-Ceramic Research and Commercialization," American Ceramic Society Bulletin, Vol. 94, #4, p 30-35 (2015).

- "Grove" = C. A. Galvin, et al. "Terracotta." Grove Art Online. Oxford Art Online. Oxford University Press, accessed July 23, 2015, subscription required.

- "Novel Ceramic Foam Is Safe And Effective Insulation". Science Daily. May 18, 2001. Retrieved November 11, 2011.

Integrating Materials Science in Ceramic Engineering

Materials science and engineering involves the process of discovery and designing of new materials. The emphasis in this process is on solids. The topics elucidated in this chapter are composite materials and nanomaterials. These topics help in broadening the existing the knowledge on ceramic engineering.

Materials Science

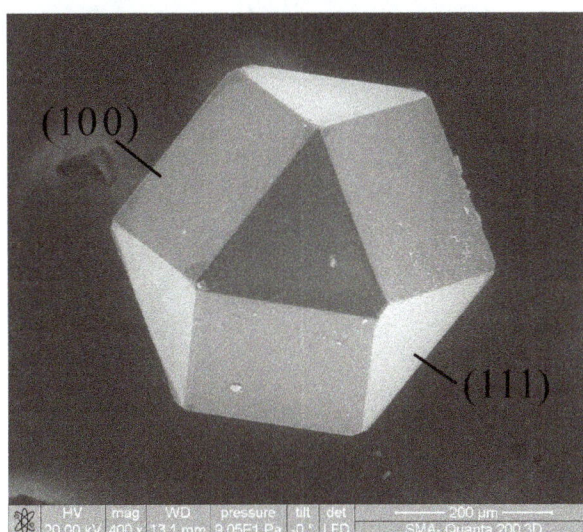

A diamond cuboctahedron showing seven crystallographic planes, imaged with scanning electron microscopy.

The interdisciplinarity field of materials science, also commonly termed materials science and engineering, involves the discovery and design of new materials, with an emphasis on solids. The intellectual origins of materials science stem from the Enlightenment, when researchers began to use analytical thinking from chemistry, physics, and engineering to understand ancient, phenomenological observations in metallurgy and mineralogy. Materials science still incorporates elements of physics, chemistry, and engineering. As such, the field was long considered by academic institutions as a sub-field of these related fields. Beginning in the 1940s, materials science began to be more widely recognized as a specific and distinct field of science and engineering, and major technical universities around the world created dedicated schools of the study.

Materials science is a syncretic discipline hybridizing metallurgy, ceramics, solid-state physics, and chemistry. It is the first example of a new academic discipline emerging by fusion rather than fission.

Many of the most pressing scientific problems humans currently face are due to the limits of the materials that are available. Thus, breakthroughs in materials science are likely to affect the future of technology significantly.

Materials scientists emphasize understanding how the history of a material (its *processing*) influences its structure, and thus the material's properties and performance. The understanding of processing-structure-properties relationships is called the § materials paradigm. This paradigm is used to advance understanding in a variety of research areas, including nanotechnology, biomaterials, and metallurgy. Materials science is also an important part of forensic engineering and failure analysis - investigating materials, products, structures or components which fail or which do not operate or function as intended, causing personal injury or damage to property. Such investigations are key to understanding, for example, the causes of various aviation accidents and incidents.

History

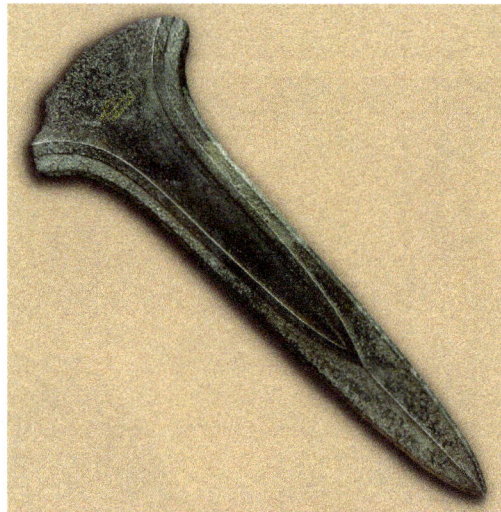

A late Bronze Age sword or dagger blade.

The material of choice of a given era is often a defining point. Phrases such as Stone Age, Bronze Age, Iron Age, and Steel Age are great examples. Originally deriving from the manufacture of ceramics and its putative derivative metallurgy, materials science is one of the oldest forms of engineering and applied science. Modern materials science evolved directly from metallurgy, which itself evolved from mining and (likely) ceramics and the use of fire. A major breakthrough in the understanding of materials occurred in the late 19th century, when the American scientist Josiah Willard Gibbs demonstrated that the thermodynamic properties related to atomic structure in various phases are related to the physical properties of a material. Important elements of modern materials science are a product of the space race: the understanding and engineering of the metallic alloys, and silica and carbon materials, used in building space vehicles enabling the exploration of space. Materials science has driven, and been driven by, the development of revolutionary technologies such as rubbers, plastics, semiconductors, and biomaterials.

Before the 1960s (and in some cases decades after), many *materials science* departments were named *metallurgy* departments, reflecting the 19th and early 20th century emphasis on metals.

The growth of materials science in the United States was catalyzed in part by the Advanced Research Projects Agency, which funded a series of university-hosted laboratories in the early 1960s "to expand the national program of basic research and training in the materials sciences." The field has since broadened to include every class of materials, including ceramics, polymers, semiconductors, magnetic materials, medical implant materials, biological materials, and nanomaterials.

Fundamentals

The materials paradigm represented in the form of a tetrahedron.

A material is defined as a substance (most often a solid, but other condensed phases can be included) that is intended to be used for certain applications. There are a myriad of materials around us—they can be found in anything from buildings to spacecraft. Materials can generally be divided into two classes: crystalline and non-crystalline. The traditional examples of materials are metals, semiconductors, ceramics and polymers. New and advanced materials that are being developed include nanomaterials and biomaterials, etc.

The basis of materials science involves studying the structure of materials, and relating them to their properties. Once a materials scientist knows about this structure-property correlation, they can then go on to study the relative performance of a material in a given application. The major determinants of the structure of a material and thus of its properties are its constituent chemical elements and the way in which it has been processed into its final form. These characteristics, taken together and related through the laws of thermodynamics and kinetics, govern a material's microstructure, and thus its properties.

Structure

As mentioned above, structure is one of the most important components of the field of materials science. Materials science examines the structure of materials from the atomic scale, all the way up to the macro scale. Characterization is the way materials scientists examine the structure of a material. This involves methods such as diffraction with X-rays, electrons, or neutrons, and various forms of spectroscopy and chemical analysis such as Raman spectroscopy, energy-dispersive spectroscopy (EDS), chromatography, thermal analysis, electron microscope analysis, etc. Structure is studied at various levels, as detailed below.

Atomic Structure

This deals with the atoms of the materials, and how they are arranged to give molecules, crystals, etc. Much of the electrical, magnetic and chemical properties of materials arise from this level of structure. The length scales involved are in angstroms. The way in which the atoms and molecules are bonded and arranged is fundamental to studying the properties and behavior of any material.

Nanostructure

Nanostructure deals with objects and structures that are in the 1—100 nm range. In many materials, atoms or molecules agglomerate together to form objects at the nanoscale. This causes many interesting electrical, magnetic, optical, and mechanical properties.

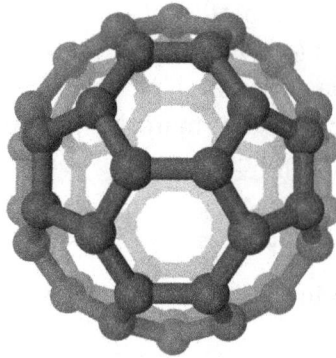

Buckminsterfullerene nanostructure.

In describing nanostructures it is necessary to differentiate between the number of dimensions on the nanoscale. Nanotextured surfaces have *one dimension* on the nanoscale, i.e., only the thickness of the surface of an object is between 0.1 and 100 nm. Nanotubes have *two dimensions* on the nanoscale, i.e., the diameter of the tube is between 0.1 and 100 nm; its length could be much greater. Finally, spherical nanoparticles have *three dimensions* on the nanoscale, i.e., the particle is between 0.1 and 100 nm in each spatial dimension. The terms nanoparticles and ultrafine particles (UFP) often are used synonymously although UFP can reach into the micrometre range. The term 'nanostructure' is often used when referring to magnetic technology. Nanoscale structure in biology is often called ultrastructure.

Materials which atoms and molecules form constituents in the nanoscale (i.e., they form nanostructure) are called nanomaterials. Nanomaterials are subject of intense research in the materials science community due to the unique properties that they exhibit.

Microstructure

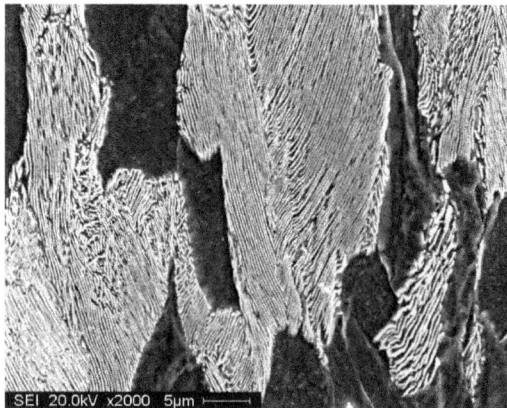

Microstructure of pearlite.

Microstructure is defined as the structure of a prepared surface or thin foil of material as revealed by a microscope above 25× magnification. It deals with objects from 100 nm to a few cm. The microstructure of a material (which can be broadly classified into metallic, polymeric, ceramic and composite) can strongly influence physical properties such as strength, toughness, ductility, hardness, corrosion resistance, high/low temperature behavior, wear resistance, and so on. Most of the traditional materials (such as metals and ceramics) are microstructured.

The manufacture of a perfect crystal of a material is physically impossible. For example, a crystalline material will contain defects such as precipitates, grain boundaries (Hall–Petch relationship), interstitial atoms, vacancies or substitutional atoms. The microstructure of materials reveals these defects, so that they can be studied.

Macro Structure

Macro structure is the appearance of a material in the scale millimeters to meters—it is the structure of the material as seen with the naked eye.

Crystallography

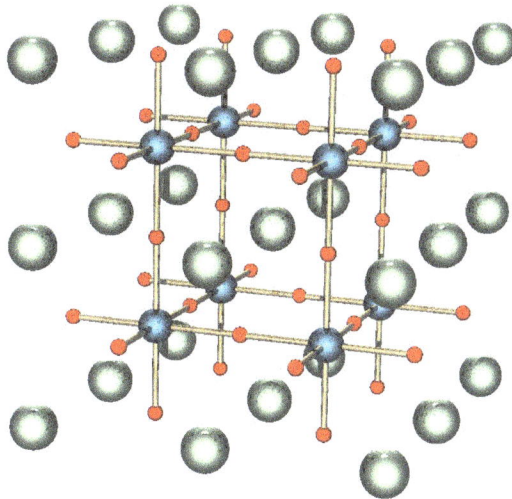

Crystal structure of a perovskite with a chemical formula ABX_3.

Crystallography is the science that examines the arrangement of atoms in crystalline solids. Crystallography is a useful tool for materials scientists. In single crystals, the effects of the crystalline arrangement of atoms is often easy to see macroscopically, because the natural shapes of crystals reflect the atomic structure. Further, physical properties are often controlled by crystalline defects. The understanding of crystal structures is an important prerequisite for understanding crystallographic defects. Mostly, materials do not occur as a single crystal, but in polycrystalline form, i.e., as an aggregate of small crystals with different orientations. Because of this, the powder diffraction method, which uses diffraction patterns of polycrystalline samples with a large number of crystals, plays an important role in structural determination. Most materials have a crystalline structure, but some important materials do not exhibit regular crystal structure. Polymers display varying degrees of crystallinity, and many are completely noncrystalline. Glass, some ceramics, and many natural materials are amorphous, not possessing any long-range order in their atomic arrangements. The study of polymers combines elements of chemical and statistical thermodynamics to give thermodynamic and mechanical, descriptions of physical properties.

Bonding

To obtain a full understanding of the material structure and how it relates to its properties, the materials scientist must study how the different atoms, ions and molecules are arranged and bonded to each other. This involves the study and use of quantum chemistry or quantum physics. Solid-state physics, solid-state chemistry and physical chemistry are also involved in the study of bonding and structure.

Properties

Materials exhibit myriad properties, including the following.

- Mechanical properties
- Chemical properties
- Electrical properties
- Thermal properties
- Optical properties
- Magnetic properties

The properties of a material determine its usability and hence its engineering application.

Synthesis and Processing

Synthesis and processing involves the creation of a material with the desired micro-nanostructure. From an engineering standpoint, a material cannot be used in industry if no economical production method for it has been developed. Thus, the processing of materials is vital to the field of materials science.

Different materials require different processing or synthesis methods. For example, the processing of metals has historically been very important and is studied under the branch of materials science named *physical metallurgy*. Also, chemical and physical methods are also used to synthesize other materials such as polymers, ceramics, thin films, etc. As of the early 21st century, new methods are being developed to synthesize nanomaterials such as graphene.

Thermodynamics

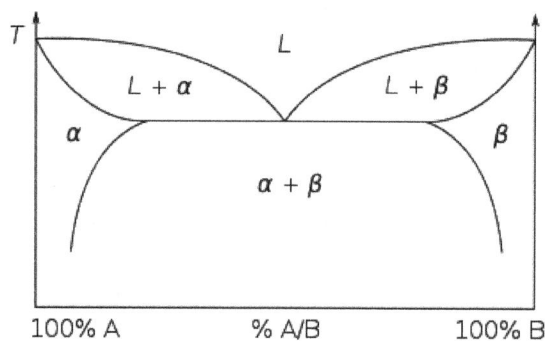

A phase diagram for a binary system displaying a eutectic point.

Thermodynamics is concerned with heat and temperature and their relation to energy and work. It defines macroscopic variables, such as internal energy, entropy, and pressure, that partly describe a body of matter or radiation. It states that the behavior of those variables is subject to general constraints, that are common to all materials, not the peculiar properties of particular materials. These general constraints are expressed in the four laws of thermodynamics. Thermodynamics describes the bulk behavior of the body, not the microscopic behaviors of the very large numbers of its microscopic constituents, such as molecules. The behavior of these microscopic particles is described by, and the laws of thermodynamics are derived from, statistical mechanics.

The study of thermodynamics is fundamental to materials science. It forms the foundation to treat general phenomena in materials science and engineering, including chemical reactions, magnetism, polarizability, and elasticity. It also helps in the understanding of phase diagrams and phase equilibrium.

Kinetics

Chemical kinetics is the study of the rates at which systems that are out of equilibrium change under the influence of various forces. When applied to materials science, it deals with how a material changes with time (moves from non-equilibrium to equilibrium state) due to application of a certain field. It details the rate of various processes evolving in materials including shape, size, composition and structure. Diffusion is important in the study of kinetics as this is the most common mechanism by which materials undergo change.

Kinetics is essential in processing of materials because, among other things, it details how the microstructure changes with application of heat.

In Research

Materials science has received much attention from researchers. In most universities, many departments ranging from physics to chemistry to chemical engineering, along with materials science departments, are involved in materials research. Research in materials science is vibrant and consists of many avenues. The following list is in no way exhaustive. It serves only to highlight certain important research areas.

Nanomaterials

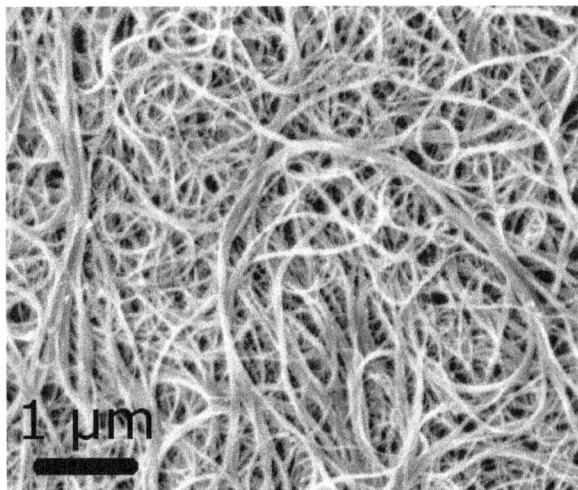

A scanning electron microscopy image of carbon nanotubes bundles

Nanomaterials describe, in principle, materials of which a single unit is sized (in at least one dimension) between 1 and 1000 nanometers (10^{-9} meter) but is usually 1—100 nm.

Nanomaterials research takes a materials science-based approach to nanotechnology, leveraging advances in materials metrology and synthesis which have been developed in support of microfabrication research. Materials with structure at the nanoscale often have unique optical, electronic, or mechanical properties.

The field of nanomaterials is loosely organized, like the traditional field of chemistry, into organic (carbon-based) nanomaterials such as fullerenes, and inorganic nanomaterials based on other elements, such as silicon. Examples of nanomaterials include fullerenes, carbon nanotubes, nanocrystals, etc.

Biomaterials

A biomaterial is any matter, surface, or construct that interacts with biological systems. As a science, *bio materials* is about fifty years old. The study of biomaterials is called *bio materials science*. It has experienced steady and strong growth over its history, with many companies investing large amounts of money into developing new products. Biomaterials science encompasses elements of medicine, biology, chemistry, tissue engineering, and materials science.

Biomaterials can be derived either from nature or synthesized in a laboratory using a variety of chemical approaches using metallic components, polymers, ceramics, or composite materials. They are often used and/or adapted for a medical application, and thus comprises whole or part of a living structure or biomedical device which performs, augments, or replaces a natural function. Such functions may be benign, like being used for a heart valve, or may be bioactive with a more interactive functionality such as hydroxylapatite coated hip implants. Biomaterials are also used everyday in dental applications, surgery, and drug delivery. For example, a construct with impregnated pharmaceutical products can be placed into the body, which permits the prolonged release of a drug over an extended period of time. A biomaterial may also be an autograft, allograft or xenograft used as an organ transplant material.

Electronic, Optical, and Magnetic

Semiconductors, metals, and ceramics are used today to form highly complex systems, such as integrated electronic circuits, optoelectronic devices, and magnetic and optical mass storage media. These materials form the basis of our modern computing world, and hence research into these materials is of vital importance.

Semiconductors are a traditional example of these types of materials. They are materials that have properties that are intermediate between conductors and insulators. Their electrical conductivities are very sensitive to impurity concentrations, and this allows for the use of doping to achieve desirable electronic properties. Hence, semiconductors form the basis of the traditional computer.

This field also includes new areas of research such as superconducting materials, spintronics, metamaterials, etc. The study of these materials involves knowledge of materials science and solid-state physics or condensed matter physics.

Computational Science and Theory

With the increase in computing power, simulating the behavior of materials has become possible. This enables materials scientists to discover properties of materials formerly unknown, as well as to design new materials. Up until now, new materials were found by time consuming trial and error processes. But, now it is hoped that computational methods could drastically reduce that time, and allow tailoring materials properties. This involves simulating materials at all length scales, using methods such as density functional theory, molecular dynamics, etc.

In Industry

Radical materials advances can drive the creation of new products or even new industries, but stable industries also employ materials scientists to make incremental improvements and troubleshoot issues with currently used materials. Industrial applications of materials science include materials design, cost-benefit tradeoffs in industrial production of materials, processing methods (casting, rolling, welding, ion implantation, crystal growth, thin-film deposition, sintering, glassblowing, etc.), and analytic methods (characterization methods such as electron microscopy, X-ray diffraction, calorimetry, nuclear microscopy (HEFIB), Rutherford backscattering, neutron diffraction, small-angle X-ray scattering (SAXS), etc.).

Besides material characterization, the material scientist or engineer also deals with extracting materials and convering them into useful forms. Thus ingot casting, foundry methods, blast furnace extraction, and electrolytic extraction are all part of the required knowledge of a materials engineer. Often the presence, absence, or variation of minute quantities of secondary elements and compounds in a bulk material will greatly affect the final properties of the materials produced. For example, steels are classified based on $1/10$ and $1/100$ weight percentages of the carbon and other alloying elements they contain. Thus, the extracting and purifying methods used to extract iron in a blast furnace can affect the quality of steel that is produced.

Ceramics and Glasses

Another application of material science is the structures of ceramics and glass, typically associated with the most brittle materials. Bonding in ceramics and glasses uses covalent and ionic-covalent types with SiO_2 (silica or sand) as a fundamental building block. Ceramics are as soft as clay or as hard as stone and concrete. Usually, they are crystalline in form. Most glasses contain a metal oxide fused with silica. At high temperatures used to prepare glass, the material is a viscous liquid. The structure of glass forms into an amorphous state upon cooling. Windowpanes and eyeglasses are important examples. Fibers of glass are also available. Scratch resistant Corning Gorilla Glass is a well-known example of the application of materials science to drastically improve the properties of common components. Diamond and carbon in its graphite form are considered to be ceramics.

Engineering ceramics are known for their stiffness and stability under high temperatures, compression and electrical stress. Alumina, silicon carbide, and tungsten carbide are made from a fine powder of their constituents in a process of sintering with a binder. Hot pressing provides higher density material. Chemical vapor deposition can place a film of a ceramic on another material. Cermets are ceramic particles containing some metals. The wear resistance of tools is derived from cemented carbides with the metal phase of cobalt and nickel typically added to modify properties.

Composites

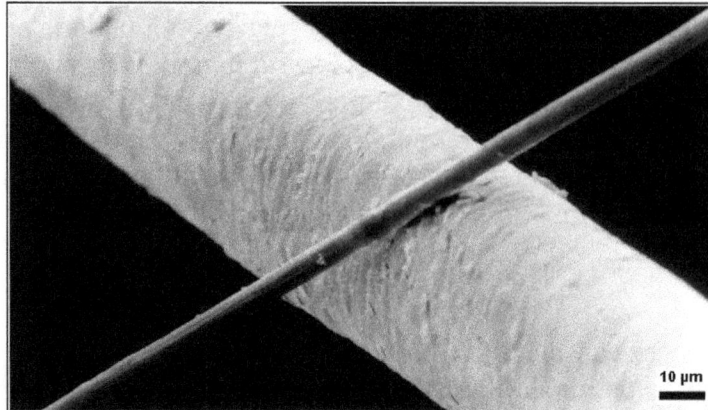

A 6 µm diameter carbon filament (running from bottom left to top right) siting atop the much larger human hair.

Filaments are commonly used for reinforcement in composite materials.

Another application of materials science in industry is making composite materials. These are structured materials composed of two or more macroscopic phases. Applications range from structural elements such as steel-reinforced concrete, to the thermal insulating tiles which play a key and integral role in NASA's Space Shuttle thermal protection system which is used to protect the surface of the shuttle from the heat of re-entry into the Earth's atmosphere. One example is reinforced Carbon-Carbon (RCC), the light gray material which withstands re-entry temperatures up to 1,510 °C (2,750 °F) and protects the Space Shuttle's wing leading edges and nose cap. RCC is a laminated composite material made from graphite rayon cloth and impregnated with a phenolic resin. After curing at high temperature in an autoclave, the laminate is pyrolized to convert the resin to carbon, impregnated with furfural alcohol in a vacuum chamber, and cured-pyrolized to convert the furfural alcohol to carbon. To provide oxidation resistance for reuse ability, the outer layers of the RCC are converted to silicon carbide.

Other examples can be seen in the "plastic" casings of television sets, cell-phones and so on. These plastic casings are usually a composite material made up of a thermoplastic matrix such as acrylonitrile butadiene styrene (ABS) in which calcium carbonate chalk, talc, glass fibers or carbon fibers have been added for added strength, bulk, or electrostatic dispersion. These additions may be termed reinforcing fibers, or dispersants, depending on their purpose.

Polymers

Polymers are chemical compounds made up of a large number of identical components linked together like chains. They are an important part of materials science. Polymers are the raw materials (the resins) used to make what are commonly called plastics and rubber. Plastics and rubber are really the final product, created after one or more polymers or additives have been added to a resin during processing, which is then shaped into a final form. Plastics which have been around, and which are in current widespread use, include polyethylene, polypropylene, polyvinyl chloride (PVC), polystyrene, nylons, polyesters, acrylics, polyurethanes, and polycarbonates and also rubbers which have been around are natural rubber, styrene butadiene rubber, chloroprene, and butadiene rubber. Plastics are generally classified as *commodity*, *specialty* and *engineering* plastics.

$$\left[\begin{array}{c} CH_3 \\ | \\ CH-CH_2 \end{array}\right]_n$$

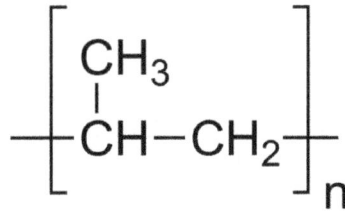

The repeating unit of the polymer polypropylene

Polyvinyl chloride (PVC) is widely used, inexpensive, and annual production quantities are large. It lends itself to an vast array of applications, from artificial leather to electrical insulation and cabling, packaging, and containers. Its fabrication and processing are simple and well-established. The versatility of PVC is due to the wide range of plasticisers and other additives that it accepts. The term "additives" in polymer science refers to the chemicals and compounds added to the polymer base to modify its material properties.

Polycarbonate would be normally considered an engineering plastic (other examples include PEEK, ABS). Such plastics are valued for their superior strengths and other special material properties. They are usually not used for disposable applications, unlike commodity plastics.

Specialty plastics are materials with unique characteristics, such as ultra-high strength, electrical conductivity, electro-fluorescence, high thermal stability, etc.

The dividing lines between the various types of plastics is not based on material but rather on their properties and applications. For example, polyethylene (PE) is a cheap, low friction polymer commonly used to make disposable bags for shopping and trash, and is considered a commodity plastic, whereas medium-density polyethylene (MDPE) is used for underground gas and water pipes, and another variety called ultra-high-molecular-weight polyethylene (UHMWPE) is an engineering plastic which is used extensively as the glide rails for industrial equipment and the low-friction socket in implanted hip joints.

Metal Alloys

Wire rope made from steel alloy.

The study of metal alloys is a significant part of materials science. Of all the metallic alloys in use today, the alloys of iron (steel, stainless steel, cast iron, tool steel, alloy steels) make up the largest proportion both by quantity and commercial value. Iron alloyed with various proportions of carbon gives low, mid and high carbon steels. An iron carbon alloy is only considered steel if the carbon level is between 0.01% and 2.00%. For the steels, the hardness and tensile strength of the steel is related to the amount of carbon present, with increasing carbon levels also leading to lower ductility and toughness. Heat treatment processes such as quenching and tempering can significantly change these properties however. Cast Iron is defined as an iron–carbon alloy with more than 2.00% but less than 6.67% carbon. Stainless steel is defined as a regular steel alloy with greater than 10% by weight alloying content of Chromium. Nickel and Molybdenum are typically also found in stainless steels.

Other significant metallic alloys are those of aluminium, titanium, copper and magnesium. Copper alloys have been known for a long time (since the Bronze Age), while the alloys of the other three metals have been relatively recently developed. Due to the chemical reactivity of these metals, the electrolytic extraction processes required were only developed relatively recently. The alloys of aluminium, titanium and magnesium are also known and valued for their high strength-to-weight ratios and, in the case of magnesium, their ability to provide electromagnetic shielding. These materials are ideal for situations where high strength-to-weight ratios are more important than bulk cost, such as in the aerospace industry and certain automotive engineering applications.

Semiconductors

The study of semiconductors is a significant part of materials science. A semiconductor is a material that has a resistivity between a metal and insulator. It's electronic properties can be greatly altered through intentionally introducing impurities, or doping. From these semiconductor materials, things such as diodes, transistors, light-emitting diodes (LEDs), and analog and digital electric circuits can be built, making them materials of interest in industry. Semiconductor devices have replaced thermionic devices (vacuum tubes) in most applications. Semiconductor devices are manufactured both as single discrete devices and as integrated circuits (ICs), which consist of a number—from a few to millions—of devices manufactured and interconnected on a single semiconductor substrate.

Of all the semiconductors in use today, silicon makes up the largest portion both by quantity and commercial value. Monocrystalline silicon is used to produce wafers used in the semiconductor and electronics industry. Second to silicon, gallium arsenide (GaAs) is the second most popular semiconductor used. Due to its higher electron mobility and saturation velocity compared to silicon, its a material of choice for high speed electronics applications. These superior properties are compelling reasons to use GaAs circuitry in mobile phones, satellite communications, microwave point-to-point links and higher frequency radar systems. Other semiconductor materials include germanium, silicon carbide, and gallium nitride and have various applications.

Relation to Other Fields

Materials science evolved—starting from the 1960s—because it was recognized that to create, discover and design new materials, one had to approach it in a unified manner. Thus, materials science and engineering emerged at the intersection of various fields such as metallurgy, solid state physics, chemistry, chemical engineering, mechanical engineering and electrical engineering.

The field is inherently interdisciplinary, and the materials scientists/engineers must be aware and make use of the methods of the physicist, chemist and engineer. The field thus, maintains close relationships with these fields. Also, many physicists, chemists and engineers also find themselves working in materials science.

The overlap between physics and materials science has led to the offshoot field of *materials physics*, which is concerned with the physical properties of materials. The approach is generally more macroscopic and applied than in condensed matter physics.

The field of materials science and engineering is important both from a scientific perspective, as well as from an engineering one. When discovering new materials, one encounters new phenomena that may not have been observed before. Hence, there is lot of science to be discovered when working with materials. Materials science also provides test for theories in condensed matter physics.

Materials are of the utmost importance for engineers, as the usage of the appropriate materials is crucial when designing systems. As a result, materials science is an increasingly important part of an engineer's education.

Emerging Technologies in Materials Science

Emerging technology	Status	Potentially marginalized technologies	Potential applications
Aerogel	Hypothetical, experiments, diffusion, early uses	Traditional insulation, glass	Improved insulation, insulative glass if it can be made clear, sleeves for oil pipelines, aerospace, high-heat & extreme cold applications
Amorphous metal	Experiments	Kevlar	Armor
Conductive polymers	Research, experiments, prototypes	Conductors	Lighter and cheaper wires, antistatic materials, organic solar cells
Femtotechnology, picotechnology	Hypothetical	Present nuclear	New materials; nuclear weapons, power
Fullerene	Experiments, diffusion	Synthetic diamond and carbon nanotubes (e.g., Buckypaper)	Programmable matter

Graphene	Hypothetical, experiments, diffusion, early uses	Silicon-based integrated circuit	Components with higher strength to weight ratios, transistors that operate at higher frequency, lower cost of display screens in mobile devices, storing hydrogen for fuel cell powered cars, filtration systems, longer-lasting and faster-charging batteries, sensors to diagnose diseases
High-temperature superconductivity	Cryogenic receiver front-end (CRFE) RF and microwave filter systems for mobile phone base stations; prototypes in dry ice; Hypothetical and experiments for higher temperatures	Copper wire, semiconductor integral circuits	No loss conductors, frictionless bearings, magnetic levitation, lossless high-capacity accumulators, electric cars, heat-free integral circuits and processors
LiTraCon	Experiments, already used to make Europe Gate	Glass	Building skyscrapers, towers, and sculptures like Europe Gate
Metamaterials	Hypothetical, experiments, diffusion	Classical optics	Microscopes, cameras, metamaterial cloaking, cloaking devices
Metal foam	Research, commercialization	Hulls	Space colonies, floating cities
Multi-function structures	Hypothetical, experiments, some prototypes, few commercial	Composite materials mostly	Wide range, e.g., self health monitoring, self healing material, morphing, ...
Nanomaterials: carbon nanotubes	Hypothetical, experiments, diffusion, early uses	Structural steel and aluminium	Stronger, lighter materials, space elevator
Programmable matter	Hypothetical, experiments	Coatings, catalysts	Wide range, e.g., claytronics, synthetic biology
Quantum dots	Research, experiments, prototypes	LCD, LED	Quantum dot laser, future use as programmable matter in display technologies (TV, projection), optical data communications (high-speed data transmission), medicine (laser scalpel)
Silicene	Hypothetical, research	Field-effect transistors	

Superalloy	Research, diffusion	Aluminum, titanium, composite materials	Aircraft jet engines
Synthetic diamond	early uses (drill bits, jewelry)	Silicon transistors	Electronics

Composite Material

A composite material (also called a composition material or shortened to composite which is the common name) is a material made from two or more constituent materials with significantly different physical or chemical properties that, when combined, produce a material with characteristics different from the individual components. The individual components remain separate and distinct within the finished structure. The new material may be preferred for many reasons: common examples include materials which are stronger, lighter, or less expensive when compared to traditional materials. More recently, researchers have also begun to actively include sensing, actuation, computation and communication into composites, which are known as Robotic Materials.

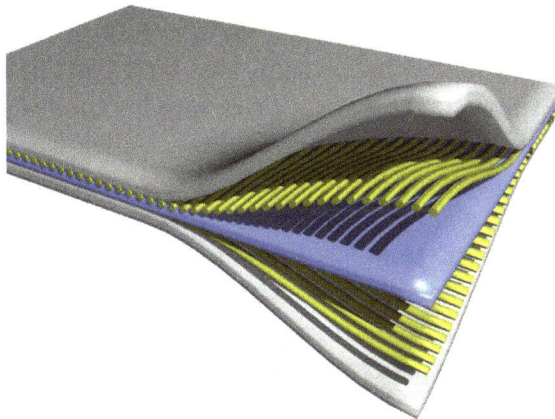

Composites are formed by combining materials together to form an overall structure with properties that differ from the sum of the individual components

Typical engineered composite materials include:

- mortars, concrete

- Reinforced plastics, such as fiber-reinforced polymer

- Metal composites

- Ceramic composites (composite ceramic and metal matrices)

Composite materials are generally used for buildings, bridges, and structures such as boat hulls, swimming pool panels, race car bodies, shower stalls, bathtubs, storage tanks, imitation granite and cultured marble sinks and countertops. The most advanced examples perform routinely on spacecraft and aircraft in demanding environments.

History

The earliest man-made composite materials were straw and mud combined to form bricks for building construction. Ancient brick-making was documented by Egyptian tomb paintings .

Wattle and daub is one of the oldest man-made composite materials, at over 6000 years old. Concrete is also a composite material, and is used more than any other man-made material in the world. As of 2006, about 7.5 billion cubic metres of concrete are made each year—more than one cubic metre for every person on Earth.

- Woody plants, both true wood from trees and such plants as palms and bamboo, yield natural composites that were used prehistorically by mankind and are still used widely in construction and scaffolding.

- Plywood 3400 BC by the Ancient Mesopotamians; gluing wood at different angles gives better properties than natural wood

- Cartonnage layers of linen or papyrus soaked in plaster dates to the First Intermediate Period of Egypt c. 2181–2055 BC and was used for death masks

- Cob (material) Mud Bricks, or Mud Walls, (using mud (clay) with straw or gravel as a binder) have been used for thousands of years.

- Concrete was described by Vitruvius, writing around 25 BC in his *Ten Books on Architecture*, distinguished types of aggregate appropriate for the preparation of lime mortars. For structural mortars, he recommended *pozzolana*, which were volcanic sands from the sandlike beds of Pozzuoli brownish-yellow-gray in colour near Naples and reddish-brown at Rome. Vitruvius specifies a ratio of 1 part lime to 3 parts pozzolana for cements used in buildings and a 1:2 ratio of lime to pulvis Puteolanus for underwater work, essentially the same ratio mixed today for concrete used at sea. Natural cement-stones, after burning, produced cements used in concretes from post-Roman times into the 20th century, with some properties superior to manufactured Portland cement.

- Papier-mâché, a composite of paper and glue, has been used for hundreds of years

- The first artificial fibre reinforced plastic was bakelite which dates to 1907, although natural polymers such as shellac predate it

- One of the most common and familiar composite is fiberglass, in which small glass fiber are embedded within a polymeric material (normally an epoxy or polyester). The glass fiber is relatively strong and stiff (but also brittle), whereas the polymer is ductile (but also weak and flexible). Thus the resulting fiberglass is relatively stiff, strong, flexible, and ductile.

Examples

Materials

Concrete is the most common artificial composite material of all and typically consists of loose stones (aggregate) held with a matrix of cement. Concrete is an inexpensive material, and will not compress or shatter even under quite a large compressive force. However, concrete cannot survive tensile loading (i.e., if stretched it will quickly break apart). Therefore, to give concrete the ability to resist being stretched, steel bars, which can resist high stretching forces, are often added to concrete to form reinforced concrete.

Concrete is a mixture of cement and aggregate, giving a robust, strong material that is very widely used.

Plywood is used widely in construction

Composite sandwich structure panel used for testing at NASA

Fibre-reinforced polymers or FRPs include carbon-fiber-reinforced polymer or CFRP, and glass-reinforced plastic or GRP. If classified by matrix then there are thermoplastic composites, short fiber thermoplastics, long fibre thermoplastics or long fibre-reinforced thermoplastics. There are numerous thermoset composites, including paper composite panels. Many advanced systems usually incorporate aramid fibre and carbon fibre in an epoxy resin matrix.

Shape memory polymer composites are high-performance composites, formulated using fibre or fabric reinforcement and shape memory polymer resin as the matrix. Since a shape memory polymer resin is used as the matrix, these composites have the ability to be easily manipulated into various configurations when they are heated above their activation temperatures and will exhibit high strength and stiffness at lower temperatures. They can also be reheated and reshaped repeatedly without losing their material properties. These composites are ideal for applications such as lightweight, rigid, deployable structures; rapid manufacturing; and dynamic reinforcement.

A. composites reinforced by particles;
B. composites reinforced by chopped strands;
C. unidirectional composites;
D. laminates;
E. fabric reinforced plastics;
F. honeycomb composite structure;

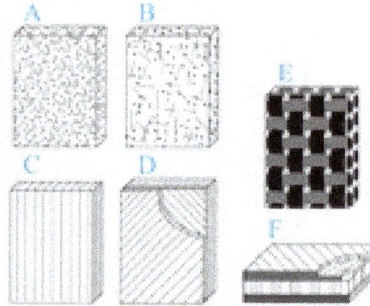

High strain composites are another type of high-performance composites that are designed to perform in a high deformation setting and are often used in deployable systems where structural flexing is advantageous. Although high strain composites exhibit many similarities to shape memory polymers, their performance is generally dependent on the fiber layout as opposed to the resin content of the matrix.

Composites can also use metal fibres reinforcing other metals, as in metal matrix composites (MMC) or ceramic matrix composites (CMC), which includes bone (hydroxyapatite reinforced with collagen fibres), cermet (ceramic and metal) and concrete. Ceramic matrix composites are built primarily for fracture toughness, not for strength.

Organic matrix/ceramic aggregate composites include asphalt concrete, polymer concrete, mastic asphalt, mastic roller hybrid, dental composite, syntactic foam and mother of pearl. Chobham armour is a special type of composite armour used in military applications.

Additionally, thermoplastic composite materials can be formulated with specific metal powders resulting in materials with a density range from 2 g/cm^3 to 11 g/cm^3 (same density as lead). The most common name for this type of material is "high gravity compound" (HGC), although "lead replacement" is also used. These materials can be used in place of traditional materials such as aluminium, stainless steel, brass, bronze, copper, lead, and even tungsten in weighting, balancing (for example, modifying the centre of gravity of a tennis racquet), vibration damping, and radiation shielding applications. High density composites are an economically viable option when certain materials are deemed hazardous and are banned (such as lead) or when secondary operations costs (such as machining, finishing, or coating) are a factor.

A sandwich-structured composite is a special class of composite material that is fabricated by attaching two thin but stiff skins to a lightweight but thick core. The core material is normally low strength material, but its higher thickness provides the sandwich composite with high bending stiffness with overall low density.

Wood is a naturally occurring composite comprising cellulose fibres in a lignin and hemicellulose matrix. Engineered wood includes a wide variety of different products such as wood fibre board, plywood, oriented strand board, wood plastic composite (recycled wood fibre in polyethylene matrix), Pykrete (sawdust in ice matrix), Plastic-impregnated or laminated paper or textiles, Arborite, Formica (plastic) and Micarta. Other engineered laminate composites, such as Mallite, use a central core of end grain balsa wood, bonded to surface skins of light alloy or GRP. These generate low-weight, high rigidity materials.

Products

Fiber-reinforced composite materials have gained popularity (despite their generally high cost) in high-performance products that need to be lightweight, yet strong enough to take harsh loading conditions such as aerospace components (tails, wings, fuselages, propellers), boat and scull hulls, bicycle frames and racing car bodies. Other uses include fishing rods, storage tanks, swimming pool panels, and baseball bats. The new Boeing 787 structure including the wings and fuselage is composed largely of composites. Composite materials are also becoming more common in the realm of orthopedic surgery.And It is the most common hockey stick material.

Carbon composite is a key material in today's launch vehicles and heat shields for the re-entry phase of spacecraft. It is widely used in solar panel substrates, antenna reflectors and yokes of spacecraft. It is also used in payload adapters, inter-stage structures and heat shields of launch vehicles. Furthermore, disk brake systems of airplanes and racing cars are using carbon/carbon material, and the composite material with carbon fibers and silicon carbide matrix has been introduced in luxury vehicles and sports cars.

In 2006, a fiber-reinforced composite pool panel was introduced for in-ground swimming pools, residential as well as commercial, as a non-corrosive alternative to galvanized steel.

In 2007, an all-composite military Humvee was introduced by TPI Composites Inc and Armor Holdings Inc, the first all-composite military vehicle. By using composites the vehicle is lighter, allowing higher payloads. In 2008, carbon fiber and DuPont Kevlar (five times stronger than steel) were combined with enhanced thermoset resins to make military transit cases by ECS Composites creating 30-percent lighter cases with high strength.

Pipes and fittings for various purpose like transportation of potable water, fire-fighting, irrigation, seawater, desalinated water, chemical and industrial waste, and sewage are now manufactured in glass reinforced plastics.

Overview

Carbon fiber composite part.

Composites are made up of individual materials referred to as constituent materials. There are two main categories of constituent materials: matrix and reinforcement. At least one portion of each type is required. The matrix material surrounds and supports the reinforcement materials by maintaining their relative positions. The reinforcements impart their special mechanical and physical properties to enhance the matrix properties. A synergism produces material properties unavailable from the individual constituent materials, while the wide variety of matrix and strengthening materials allows the designer of the product or structure to choose an optimum combination.

Engineered composite materials must be formed to shape. The matrix material can be introduced to the reinforcement before or after the reinforcement material is placed into the mould cavity or onto the mould surface. The matrix material experiences a melding event, after which the part shape is essentially set. Depending upon the nature of the matrix material, this melding event can occur in various ways such as chemical polymerization or solidification from the melted state.

A variety of moulding methods can be used according to the end-item design requirements. The principal factors impacting the methodology are the natures of the chosen matrix and reinforcement materials. Another important factor is the gross quantity of material to be produced. Large quantities can be used to justify high capital expenditures for rapid and automated manufacturing technology. Small production quantities are accommodated with lower capital expenditures but higher labour and tooling costs at a correspondingly slower rate.

Many commercially produced composites use a polymer matrix material often called a resin solution. There are many different polymers available depending upon the starting raw ingredients. There are several broad categories, each with numerous variations. The most common are known as polyester, vinyl ester, epoxy, phenolic, polyimide, polyamide, polypropylene, PEEK, and others. The reinforcement materials are often fibres but also commonly ground minerals. The various methods described below have been developed to reduce the resin content of the final product, or the fibre content is increased. As a rule of thumb, lay up results in a product containing 60% resin and 40% fibre, whereas vacuum infusion gives a final product with 40% resin and 60% fiber content. The strength of the product is greatly dependent on this ratio.

Martin Hubbe and Lucian A Lucia consider wood to be a natural composite of cellulose fibres in a matrix of lignin.

Constituents

Organic

Polymers are common matrices (especially used for fiber reinforced plastics). Road surfaces are often made from asphalt concrete which uses bitumen as a matrix. Mud (wattle and daub) has seen extensive use. Typically, most common polymer-based composite materials, including fiberglass, carbon fiber, and Kevlar, include at least two parts, the substrate and the resin.

Polyester resin tends to have yellowish tint, and is suitable for most backyard projects. Its weaknesses are that it is UV sensitive and can tend to degrade over time, and thus generally is also coated to help preserve it. It is often used in the making of surfboards and for marine applications. Its hardener is a peroxide, often MEKP (methyl ethyl ketone peroxide). When the peroxide is mixed with the resin, it decomposes to generate free radicals, which initiate the curing reaction. Hardeners in these systems are commonly called catalysts, but since they do not re-appear unchanged at the end of the reaction, they do not fit the strictest chemical definition of a catalyst.

Vinylester resin tends to have a purplish to bluish to greenish tint. This resin has lower viscosity than polyester resin, and is more transparent. This resin is often billed as being fuel resistant, but will melt in contact with gasoline. This resin tends to be more resistant over time to degradation than polyester resin, and is more flexible. It uses the same hardeners as polyester resin (at a similar mix ratio) and the cost is approximately the same.

Epoxy resin is almost totally transparent when cured. In the aerospace industry, epoxy is used as a structural matrix material or as a structural glue.

Shape memory polymer (SMP) resins have varying visual characteristics depending on their formulation. These resins may be epoxy-based, which can be used for auto body and outdoor equipment repairs; cyanate-ester-based, which are used in space applications; and acrylate-based, which can be used in very cold temperature applications, such as for sensors that indicate whether perishable goods have warmed above a certain maximum temperature. These resins are unique in that their shape can be repeatedly changed by heating above their glass transition temperature (T_g). When heated, they become flexible and elastic, allowing for easy configuration. Once they are cooled, they will maintain their new shape. The resins will return to their original shapes when they are reheated above their T_g. The advantage of shape memory polymer resins is that they can be shaped and reshaped repeatedly without losing their material properties. These resins can be used in fabricating shape memory composites.

Inorganic

Cement (concrete), metals, ceramics, and sometimes glasses are employed. Unusual matrices such as ice are sometime proposed as in pykecrete.

Reinforcements

Fiber

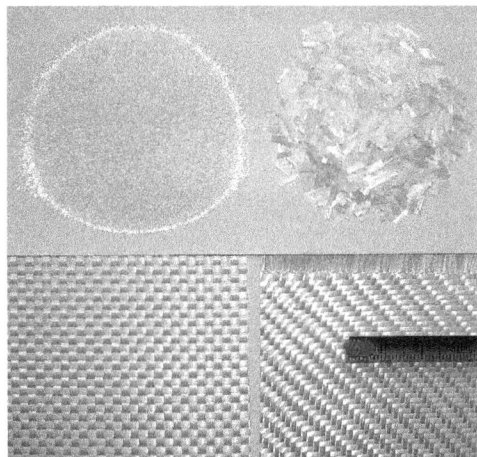

Differences in the way the fibers are laid out give different strengths and ease of manufacture

Reinforcement usually adds rigidity and greatly impedes crack propagation. Thin fibers can have very high strength, and provided they are mechanically well attached to the matrix they can greatly improve the composite's overall properties.

Fiber-reinforced composite materials can be divided into two main categories normally referred to as short fiber-reinforced materials and continuous fiber-reinforced materials. Continuous reinforced materials will often constitute a layered or laminated structure. The woven and continuous fibre styles are typically available in a variety of forms, being pre-impregnated with the given matrix (resin), dry, uni-directional tapes of various widths, plain weave, harness satins, braided, and stitched.

The short and long fibers are typically employed in compression moulding and sheet moulding operations. These come in the form of flakes, chips, and random mate (which can also be made from a continuous fibre laid in random fashion until the desired thickness of the ply / laminate is achieved).

Common fibers used for reinforcement include glass fibers, carbon fibers, cellulose (wood/paper fiber and straw) and high strength polymers for example aramid. Silicon carbide fibers are used for some high temperature applications.

Other Reinforcement

Concrete uses aggregate, and reinforced concrete additionally uses steel bars (rebar) to tension the concrete. Steel mesh or wires are also used in some glass and plastic products.

Cores

Many composite layup designs also include a co-curing or post-curing of the prepreg with various other media, such as honeycomb or foam. This is commonly called a sandwich structure. This is a more common layup for the manufacture of radomes, doors, cowlings, or non-structural parts.

Open- and closed-cell-structured foams like polyvinylchloride, polyurethane, polyethylene or polystyrene foams, balsa wood, syntactic foams, and honeycombs are commonly used core materials. Open- and closed-cell metal foam can also be used as core materials. Recently, 3D graphene structures (also called graphene foam) have also been employed as core structures.A recent review by Khurram and Xu et al., have provided the summary of the state-of-the-art techniques for fabrication of the 3D structure of graphene, and the examples of the use of these foam like structures as a core for their respective polymer composites.

Fabrication Methods

Fabrication of composite materials is accomplished by a wide variety of techniques, including:

- Advanced fiber placement (Automated fiber placement)
- Tailored fiber placement
- Fiberglass spray lay-up process
- Filament winding

- Lanxide process

- Tufting

- Z-pinning

Composite fabrication usually involves wetting, mixing or saturating the reinforcement with the matrix, and then causing the matrix to bind together (with heat or a chemical reaction) into a rigid structure. The operation is usually done in an open or closed forming mold, but the order and ways of introducing the ingredients varies considerably.

Mold Overview

Within a mold, the reinforcing and matrix materials are combined, compacted, and cured (processed) to undergo a melding event. After the melding event, the part shape is essentially set, although it can deform under certain process conditions. For a thermoset polymeric matrix material, the melding event is a curing reaction that is initiated by the application of additional heat or chemical reactivity such as an organic peroxide. For a thermoplastic polymeric matrix material, the melding event is a solidification from the melted state. For a metal matrix material such as titanium foil, the melding event is a fusing at high pressure and a temperature near the melting point.

For many moulding methods, it is convenient to refer to one mould piece as a "lower" mould and another mould piece as an "upper" mould. Lower and upper refer to the different faces of the moulded panel, not the mould's configuration in space. In this convention, there is always a lower mould, and sometimes an upper mould. Part construction begins by applying materials to the lower mould. Lower mould and upper mould are more generalized descriptors than more common and specific terms such as male side, female side, a-side, b-side, tool side, bowl, hat, mandrel, etc. Continuous manufacturing uses a different nomenclature.

The moulded product is often referred to as a panel. For certain geometries and material combinations, it can be referred to as a casting. For certain continuous processes, it can be referred to as a profile.

Vacuum Bag Moulding

Vacuum bag moulding uses a flexible film to enclose the part and seal it from outside air. Vacuum bag material is available in a tube shape or a sheet of material. A vacuum is then drawn on the vacuum bag and atmospheric pressure compresses the part during the cure. When a tube shaped bag is used, the entire part can be enclosed within the bag. When using sheet bagging materials, the edges of the vacuum bag are sealed against the edges of the mould surface to enclose the part against an air-tight mould. When bagged in this way, the lower mould is a rigid structure and the upper surface of the part is formed by the flexible membrane vacuum bag. The flexible membrane can be a reusable silicone material or an extruded polymer film. After sealing the part inside the vacuum bag, a vacuum is drawn on the part (and held) during cure. This process can be performed at either ambient or elevated temperature with ambient atmospheric pressure acting upon the vacuum bag. A vacuum pump is typically used to draw a vacuum. An economical method of drawing a vacuum is with a venturi vacuum and air compressor.

A vacuum bag is a bag made of strong rubber-coated fabric or a polymer film used to compress the part during cure or hardening. In some applications the bag encloses the entire material, or in other applications a mold is used to form one face of the laminate with the bag being a single layer to seal to the outer edge of the mold face. When using a tube shaped bag, the ends of the bag are sealed and the air is drawn out of the bag through a nipple using a vacuum pump. As a result, uniform pressure approaching one atmosphere is applied to the surfaces of the object inside the bag, holding parts together while the adhesive cures. The entire bag may be placed in a temperature-controlled oven, oil bath or water bath and gently heated to accelerate curing.

Vacuum bagging is widely used in the composites industry as well. Carbon fiber fabric and fiberglass, along with resins and epoxies are common materials laminated together with a vacuum bag operation.

Woodworking applications

In commercial woodworking facilities, vacuum bags are used to laminate curved and irregular shaped workpieces.

Typically, polyurethane or vinyl materials are used to make the bag. A tube shaped bag is open at both ends. The piece, or pieces to be glued are placed into the bag and the ends sealed. One method of sealing the open ends of the bag is by placing a clamp on each end of the bag. A plastic rod is laid across the end of the bag, the bag is then folded over the rod. A plastic sleeve with an opening in it, is then snapped over the rod. This procedure forms a seal at both ends of the bag, when the vacuum is ready to be drawn.

A "platen" is sometimes used inside the bag for the piece being glued to lie on. The platen has a series of small slots cut into it, to allow the air under it to be evacuated. The platen must have rounded edges and corners to prevent the vacuum from tearing the bag.

When a curved part is to be glued in a vacuum bag, it is important that the pieces being glued be placed over a solidly built form, or have an air bladder placed under the form. This air bladder has access to "free air" outside the bag. It is used to create an equal pressure under the form, preventing it from being crushed.

Pressure Bag Moulding

This process is related to vacuum bag molding in exactly the same way as it sounds. A solid female mold is used along with a flexible male mold. The reinforcement is placed inside the female mold with just enough resin to allow the fabric to stick in place (wet lay up). A measured amount of resin is then liberally brushed indiscriminately into the mold and the mold is then clamped to a machine that contains the male flexible mold. The flexible male membrane is then inflated with heated compressed air or possibly steam. The female mold can also be heated. Excess resin is forced out along with trapped air. This process is extensively used in the production of composite helmets due to the lower cost of unskilled labor. Cycle times for a helmet bag moulding machine vary from 20 to 45 minutes, but the finished shells require no further curing if the molds are heated.

Autoclave Moulding

A process using a two-sided mould set that forms both surfaces of the panel. On the lower side is a rigid mould and on the upper side is a flexible membrane made from silicone or an extruded polymer film such as nylon. Reinforcement materials can be placed manually or robotically. They include continuous fibre forms fashioned into textile constructions. Most often, they are pre-impregnated with the resin in the form of prepreg fabrics or unidirectional tapes. In some instances, a resin film is placed upon the lower mould and dry reinforcement is placed above. The upper mould is installed and vacuum is applied to the mould cavity. The assembly is placed into an autoclave. This process is generally performed at both elevated pressure and elevated temperature. The use of elevated pressure facilitates a high fibre volume fraction and low void content for maximum structural efficiency.

Resin Transfer Moulding (RTM)

RTM is a process using a rigid two-sided mould set that forms both surfaces of the panel. The mould is typically constructed from aluminum or steel, but composite molds are sometimes used. The two sides fit together to produce a mould cavity. The distinguishing feature of resin transfer moulding is that the reinforcement materials are placed into this cavity and the mould set is closed prior to the introduction of matrix material. Resin transfer moulding includes numerous varieties which differ in the mechanics of how the resin is introduced to the reinforcement in the mould cavity. These variations include everything from the RTM methods used in out of autoclave composite manufacturing for high-tech aerospace components to vacuum infusion to vacuum assisted resin transfer moulding (VARTM). This process can be performed at either ambient or elevated temperature.

Other Fabrication Methods

Other types of fabrication include press moulding, transfer moulding, pultrusion moulding, filament winding, casting, centrifugal casting, continuous casting and slip forming. There are also forming capabilities including CNC filament winding, vacuum infusion, wet lay-up, compression moulding, and thermoplastic moulding, to name a few. The use of curing ovens and paint booths is also needed for some projects.

Finishing Methods

The finishing of the composite parts is also critical in the final design. Many of these finishes will include rain-erosion coatings or polyurethane coatings.

Tooling

The mold and mold inserts are referred to as "tooling." The mold/tooling can be constructed from a variety of materials. Tooling materials include invar, steel, aluminium, reinforced silicone rubber, nickel, and carbon fiber. Selection of the tooling material is typically based on, but not limited to, the coefficient of thermal expansion, expected number of cycles, end item tolerance, desired or required surface condition, method of cure, glass transition temperature of the material being moulded, moulding method, matrix, cost and a variety of other considerations.

Physical Properties

The physical properties of composite materials are generally not isotropic (independent of direction of applied force) in nature, but rather are typically anisotropic (different depending on the direction of the applied force or load). For instance, the stiffness of a composite panel will often depend upon the orientation of the applied forces and/or moments. Panel stiffness is also dependent on the design of the panel. For instance, the fibre reinforcement and matrix used, the method of panel build, thermoset versus thermoplastic, type of weave, and orientation of fibre axis to the primary force.

In contrast, isotropic materials (for example, aluminium or steel), in standard wrought forms, typically have the same stiffness regardless of the directional orientation of the applied forces and/or moments.

The relationship between forces/moments and strains/curvatures for an isotropic material can be described with the following material properties: Young's Modulus, the shear Modulus and the Poisson's ratio, in relatively simple mathematical relationships. For the anisotropic material, it requires the mathematics of a second order tensor and up to 21 material property constants. For the special case of orthogonal isotropy, there are three different material property constants for each of Young's Modulus, Shear Modulus and Poisson's ratio—a total of 9 constants to describe the relationship between forces/moments and strains/curvatures.

Techniques that take advantage of the anisotropic properties of the materials include mortise and tenon joints (in natural composites such as wood) and Pi Joints in synthetic composites.

Failure

Shock, impact, or repeated cyclic stresses can cause the laminate to separate at the interface between two layers, a condition known as delamination. Individual fibres can separate from the matrix e.g. fibre pull-out.

Composites can fail on the microscopic or macroscopic scale. Compression failures can occur at both the macro scale or at each individual reinforcing fiber in compression buckling. Tension failures can be net section failures of the part or degradation of the composite at a microscopic scale where one or more of the layers in the composite fail in tension of the matrix or failure of the bond between the matrix and fibers.

Some composites are brittle and have little reserve strength beyond the initial onset of failure while others may have large deformations and have reserve energy absorbing capacity past the onset of damage. The variations in fibers and matrices that are available and the mixtures that can be made with blends leave a very broad range of properties that can be designed into a composite structure. The best known failure of a brittle ceramic matrix composite occurred when the carbon-carbon composite tile on the leading edge of the wing of the Space Shuttle Columbia fractured when impacted during take-off. It led to catastrophic break-up of the vehicle when it re-entered the Earth's atmosphere on 1 February 2003.

Compared to metals, composites have relatively poor bearing strength.

Testing

To aid in predicting and preventing failures, composites are tested before and after construction. Pre-construction testing may use finite element analysis (FEA) for ply-by-ply analysis of curved surfaces and predicting wrinkling, crimping and dimpling of composites. Materials may be tested during manufacturing and after construction through several nondestructive methods including ultrasonics, thermography, shearography and X-ray radiography, and laser bond inspection for NDT of relative bond strength integrity in a localized area.

Nanomaterials

Nanomaterials describe, in principle, materials of which a single unit is sized (in at least one dimension) between 1 and 1000 nanometres (10^{-9} meter) but is usually 1—100 nm (the usual definition of nanoscale).

Nanomaterials research takes a materials science-based approach to nanotechnology, leveraging advances in materials metrology and synthesis which have been developed in support of microfabrication research. Materials with structure at the nanoscale often have unique optical, electronic, or mechanical properties.

Nanomaterials are slowly becoming commercialized and beginning to emerge as commodities.

Types

Natural Nanomaterials

Biological systems often feature natural, functional nanomaterials. The structure of foraminifera (mainly chalk) and viruses (protein, capsid), the wax crystals covering a lotus or nasturtium leaf, spider and spider-mite silk, the blue hue of tarantulas, the "spatulae" on the bottom of gecko feet, some butterfly wing scales, natural colloids (milk, blood), horny materials (skin, claws, beaks, feathers, horns, hair), paper, cotton, nacre, corals, and even our own bone matrix are all natural *organic* nanomaterials.

Natural *inorganic* nanomaterials occur through crystal growth in the diverse chemical conditions of the Earth's crust. For example, clays display complex nanostructures due to anisotropy of their underlying crystal structure, and volcanic activity can give rise to opals, which are an instance of a naturally occurring photonic crystals due to their nanoscale structure. Fires represent particularly complex reactions and can produce pigments, cement, fumed silica etc.

| Viral capsid | "Lotus effect", hydrophobic effect with self-cleaning ability | Close-up of the underside of a gecko's foot as it walks on a glass wall. (spatula: $200 \times 10\text{-}15$ nm). | SEM micrograph of a butterfly wing scale ($\times$ 5000) |

Peacock feather (detail)	Brazilian Crystal Opal. The play of color is caused by the interference and diffraction of light between silica spheres (150 - 300 nm in diameter).	Lycurgus Cup, glass, 4th century, Roman. Nanoparticles (70 nm) of gold and silver, dispersed in colloidal form, are responsible for the dichroic effect (red/green).	Blue hue of a species of tarantula (450 nm ± 20 nm)

Fullerenes

The fullerenes are a class of allotropes of carbon which conceptually are graphene sheets rolled into tubes or spheres. These include the carbon nanotubes (or silicon nanotubes) which are of interest both because of their mechanical strength and also because of their electrical properties.

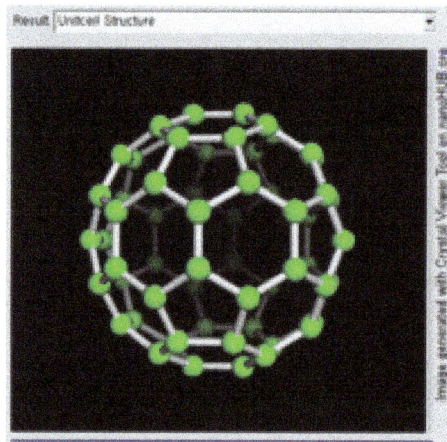

Rotating view of C_{60}, one kind of fullerene.

The first fullerene molecule to be discovered, and the family's namesake, buckminsterfullerene (C_{60}), was prepared in 1985 by Richard Smalley, Robert Curl, James Heath, Sean O'Brien, and Harold Kroto at Rice University. The name was a homage to Buckminster Fuller, whose geodesic domes it resembles. Fullerenes have since been found to occur in nature. More recently, fullerenes have been detected in outer space.

For the past decade, the chemical and physical properties of fullerenes have been a hot topic in the field of research and development, and are likely to continue to be for a long time. In April 2003, fullerenes were under study for potential medicinal use: binding specific antibiotics to the structure of resistant bacteria and even target certain types of cancer cells such as melanoma. The October 2005 issue of Chemistry and Biology contains an article describing the use of fullerenes as light-activated antimicrobial agents. In the field of nanotechnology, heat resistance and superconductivity are among the properties attracting intense research.

A common method used to produce fullerenes is to send a large current between two nearby graphite electrodes in an inert atmosphere. The resulting carbon plasma arc between the electrodes cools into sooty residue from which many fullerenes can be isolated.

There are many calculations that have been done using ab-initio Quantum Methods applied to fullerenes. By DFT and TDDFT methods one can obtain IR, Raman and UV spectra. Results of such calculations can be compared with experimental results.

Graphene Nanostructures

2D materials are crystalline materials consisting of a two-dimensional single layer of atoms. The most important representative graphene was discovered in 2004. Other 2D materials based on other elements have since been reported.

Box-shaped graphene (BSG) nanostructure is an example of 3D nanomaterial. BSG nanostructure has appeared after mechanical cleavage of pyrolytic graphite. This nanostructure is a multi-layer system of parallel hollow nanochannels located along the surface and having quadrangular cross-section. The thickness of the channel walls is approximately equal to 1 nm. The typical width of channel facets makes about 25 nm.

Nanoparticles

Inorganic nanomaterials, (e.g. quantum dots, nanowires and nanorods) because of their interesting optical and electrical properties, could be used in optoelectronics. Furthermore, the optical and electronic properties of nanomaterials which depend on their size and shape can be tuned via synthetic techniques. There are the possibilities to use those materials in organic material based optoelectronic devices such as Organic solar cells, OLEDs etc. The operating principles of such devices are governed by photoinduced processes like electron transfer and energy transfer. The performance of the devices depends on the efficiency of the photoinduced process responsible for their functioning. Therefore, better understanding of those photoinduced processes in organic/inorganic nanomaterial composite systems is necessary in order to use them in organic optoelectronic devices.

Nanoparticles or nanocrystals made of metals, semiconductors, or oxides are of particular interest for their mechanical, electrical, magnetic, optical, chemical and other properties. Nanoparticles have been used as quantum dots and as chemical catalysts such as nanomaterial-based catalysts. Recently, a range of nanoparticles are extensively investigated for biomedical applications including tissue engineering, drug delivery, biosensor.

Nanoparticles are of great scientific interest as they are effectively a bridge between bulk materials and atomic or molecular structures. A bulk material should have constant physical properties regardless of its size, but at the nano-scale this is often not the case. Size-dependent properties are observed such as quantum confinement in semiconductor particles, surface plasmon resonance in some metal particles and superparamagnetism in magnetic materials.

Nanoparticles exhibit a number of special properties relative to bulk material. For example, the bending of bulk copper (wire, ribbon, etc.) occurs with movement of copper atoms/clusters at about the 50 nm scale. Copper nanoparticles smaller than 50 nm are considered super hard materials that do not exhibit the same malleability and ductility as bulk copper. The change in properties is not always desirable. Ferroelectric materials smaller than 10 nm can switch their magnetisation direction using room temperature thermal energy, thus making them useless for memory storage. Suspensions of nanoparticles are possible because the interaction of the particle surface with the solvent is strong enough to overcome differences in density, which usually result in a material either sinking or floating in a liquid. Nanoparticles often have unexpected visual properties because they are small enough to confine their electrons and produce quantum effects. For example, gold nanoparticles appear deep red to black in solution.

The often very high surface area to volume ratio of nanoparticles provides a tremendous driving force for diffusion, especially at elevated temperatures. Sintering is possible at lower temperatures and over shorter durations than for larger particles. This theoretically does not affect the density of the final product, though flow difficulties and the tendency of nanoparticles to agglomerate do complicate matters. The surface effects of nanoparticles also reduces the incipient melting temperature.

Nanozymes

Nanozymes are nanomaterials with enzyme-like characteristics. They are an emerging type of artificial enzyme, which have been used for wide applications in such as biosensing, bioimaging, tumor diagnosis and therapy, antibiofouling, etc.

Synthesis

The goal of any synthetic method for nanomaterials is to yield a material that exhibits properties that are a result of their characteristic length scale being in the nanometer range (~1 – 100 nm). Accordingly, the synthetic method should exhibit control of size in this range so that one property or another can be attained. Often the methods are divided into two main types "Bottom Up" and "Top Down."

Bottom Up Methods

Bottom up methods involve the assembly of atoms or molecules into nanostructured arrays. In these methods the raw material sources can be in the form of gases, liquids or solids. The latter requiring some sort of disassembly prior to their incorporation onto a nanostructure. Bottom methods generally fall into two categories: chaotic and controlled.

Chaotic Processes

Chaotic processes involve elevating the constituent atoms or molecules to a chaotic state and then suddenly changing the conditions so as to make that state unstable. Through the clever manipulation of any number of parameters, products form largely as a result of the insuring kinetics. The collapse from the chaotic state can be difficult or impossible to control and so ensemble statistics often govern the resulting size distribution and average size. Accordingly, nanoparticle formation is controlled through manipulation of the end state of the products.

Examples of Chaotic Processes are: Laser ablation, Exploding wire, Arc, Flame pyrolysis, Combustion, Precipitation synthesis techniques.

Controlled Processes

Controlled Processes involve the controlled delivery of the constituent atoms or molecules to the site(s) of nanoparticle formation such that the nanoparticle can grow to a prescribed sizes in a controlled manner. Generally the state of the constituent atoms or molecules are never far from that needed for nanoparticle formation. Accordingly, nanoparticle formation is controlled through the control of the state of the reactants.

Examples of controlled processes are self-limiting growth solution, self-limited chemical vapor deposition, shaped pulse femtosecond laser techniques, and molecular beam epitaxy.

Characterization

Novel effects can occur in materials when structures are formed with sizes comparable to any one of many possible length scales, such as the de Broglie wavelength of electrons, or the optical wavelengths of high energy photons. In these cases quantum mechanical effects can dominate material properties. One example is quantum confinement where the electronic properties of solids are altered with great reductions in particle size. The optical properties of nanoparticles, e.g. fluorescence, also become a function of the particle diameter. This effect does not come into play by going from macrosocopic to micrometer dimensions, but becomes pronounced when the nanometer scale is reached.

In addition to optical and electronic properties, the novel mechanical properties of many nanomaterials is the subject of nanomechanics research. When added to a bulk material, nanoparticles can strongly influence the mechanical properties of the material, such as the stiffness or elasticity. For example, traditional polymers can be reinforced by nanoparticles (such as carbon nanotubes) resulting in novel materials which can be used as lightweight replacements for metals. Such composite materials may enable a weight reduction accompanied by an increase in stability and improved functionality.

Finally, nanostructured materials with small particle size such as zeolites, and asbestos, are used as catalysts in a wide range of critical industrial chemical reactions. The further development of such catalysts can form the basis of more efficient, environmentally friendly chemical processes.

The first observations and size measurements of nano-particles were made during the first decade of the 20th century. Zsigmondy made detailed studies of gold sols and other nanomaterials with sizes down to 10 nm and less. He published a book in 1914. He used an ultramicroscope that employs a *dark field* method for seeing particles with sizes much less than light wavelength.

There are traditional techniques developed during 20th century in Interface and Colloid Science for characterizing nanomaterials. These are widely used for *first generation* passive nanomaterials specified in the next section.

These methods include several different techniques for characterizing particle size distribution. This characterization is imperative because many materials that are expected to be nano-sized are actually aggregated in solutions. Some of methods are based on light scattering. Others apply ultrasound, such as ultrasound attenuation spectroscopy for testing concentrated nano-dispersions and microemulsions.

There is also a group of traditional techniques for characterizing surface charge or zeta potential of nano-particles in solutions. This information is required for proper system stabilzation, preventing its aggregation or flocculation. These methods include microelectrophoresis, electrophoretic light scattering and electroacoustics. The last one, for instance colloid vibration current method is suitable for characterizing concentrated systems.

Uniformity

The chemical processing and synthesis of high performance technological components for the private, industrial and military sectors requires the use of high purity ceramics, polymers, glass-ceramics and material composites. In condensed bodies formed from fine powders, the irregular sizes and shapes of nanoparticles in a typical powder often lead to non-uniform packing morphologies that result in packing density variations in the powder compact.

Uncontrolled agglomeration of powders due to attractive van der Waals forces can also give rise to in microstructural inhomogeneities. Differential stresses that develop as a result of non-uniform drying shrinkage are directly related to the rate at which the solvent can be removed, and thus highly dependent upon the distribution of porosity. Such stresses have been associated with a plastic-to-brittle transition in consolidated bodies, and can yield to crack propagation in the unfired body if not relieved.

In addition, any fluctuations in packing density in the compact as it is prepared for the kiln are often amplified during the sintering process, yielding inhomogeneous densification. Some pores and other structural defects associated with density variations have been shown to play a detrimental role in the sintering process by growing and thus limiting end-point densities. Differential stresses arising from inhomogeneous densification have also been shown to result in the propagation of internal cracks, thus becoming the strength-controlling flaws.

It would therefore appear desirable to process a material in such a way that it is physically uniform with regard to the distribution of components and porosity, rather than using particle size distributions which will maximize the green density. The containment of a uniformly dispersed assembly of strongly interacting particles in suspension requires total control over particle-particle interactions. It should be noted here that a number of dispersants such as ammonium citrate (aqueous) and imidazoline or oleyl alcohol (nonaqueous) are promising solutions as possible additives for enhanced dispersion and deagglomeration. Monodisperse nanoparticles and colloids provide this potential.

Monodisperse powders of colloidal silica, for example, may therefore be stabilized sufficiently to ensure a high degree of order in the colloidal crystal or polycrystalline colloidal solid which results from aggregation. The degree of order appears to be limited by the time and space allowed for longer-range correlations to be established. Such defective polycrystalline colloidal structures would appear to be the basic elements of sub-micrometer colloidal materials science, and, therefore, provide the first step in developing a more rigorous understanding of the mechanisms involved in microstructural evolution in high performance materials and components.

Legal Definition

On 18 October 2011, the European Commission adopted the following definition of a nanomaterial: "A natural, incidental or manufactured material containing particles, in an unbound state or as an aggregate or as an agglomerate and where, for 50% or more of the particles in the number size distribution, one or more external dimensions is in the size range 1 nm – 100 nm. In specific cases and where warranted by concerns for the environment, health, safety or competitiveness the number size distribution threshold of 50% may be replaced by a threshold between 1 and 50%."

However, this differs from the definition adopted by the International Organization for Standardization (ISO), which is: "Material with any external dimension in the nanoscale or having internal structure in the nanoscale."

"Nanoscale" is, in turn, defined as: "Size range from approximately 1 nm to 100 nm."

It is not currently known which of these, if any, will prevail in courts of law.

Health Impact

While nanomaterials and nanotechnologies are expected to yield numerous health and health care advances, such as more targeted methods of delivering drugs, new cancer therapies, and methods of early detection of diseases, they also may have unwanted effects. Increased rate of absorption is the main concern associated with manufactured nanoparticles.

When materials are made into nanoparticles, their surface area to volume ratio increases. The greater specific surface area (surface area per unit weight) may lead to increased rate of absorption through the skin, lungs, or digestive tract and may cause unwanted effects to the lungs as well as other organs. However, the particles must be absorbed in sufficient quantities in order to pose health risks.

Nanoparticles created adventitiously (e.g., through the rubbing of prostheses) have long been known to be a health hazard, but as the use of nanomaterials increases worldwide, concerns for worker and user safety are mounting. To address such concerns, the Swedish Karolinska Institute conducted a study in which various nanoparticles were introduced to human lung epithelial cells. The results, released in 2008, showed that iron oxide nanoparticles caused little DNA damage and were non-toxic. Zinc oxide nanoparticles were slightly worse. Titanium dioxide caused only DNA damage. Carbon nanotubes caused DNA damage at low levels. Copper oxide was found to be the worst offender, and was the only nanomaterial identified by the researchers as a clear health risk. Though nanomaterials are not confirmed as a health risk to workers who produce them, NIOSH recommends that exposure precautions and personal protective equipment be used to protect workers until the risks of nanomaterial manufacture are better understood.

Occupational Safety

Beyond normal chemical safety practices such as using personal protective equipment, The United States National Institute for Occupational Safety and Health recommends the use of HEPA filtration on local exhaust ventilation such as fume hoods. General laboratory ventilation is often not sufficient to effectively clear nanomaterials released into the general room air over a 30-minute period; therefore, researchers should leave the hood fan on even after synthesis is complete. Nanomaterials in dry powder form have the most risk of airborne dust generation and dermal contact, followed by liquid suspensions and then by nanomaterials embedded in a solid matrix.

Nanoparticles behave differently than other similarly sized particles. It is therefore necessary to develop specialized approaches to testing and monitoring their effects on human health and on the environment. The OECD Chemicals Committee has established the Working Party on Manufactured Nanomaterials to address this issue and to study the practices of OECD member countries in regards to nanomaterial safety.

References

- "General Safe Practices for Working with Engineered Nanomaterials in Research Laboratories". National Institute of Occupational Safety and Health. May 2012. pp. 6–8. Retrieved 2016-07-15.

- Khurram, Shehzad; Xu, Yang; Chao, Gao; Xianfeng, Duan (2016). "Three-dimensional macro-structures of two-dimensional nanomaterials". Chemical Society Reviews. doi:10.1039/C6CS00218H.

- Kerativitayanan, P; Carrow, JK; Gaharwar, AK (26 May 2015). "Nanomaterials for Engineering Stem Cell Responses.". Advanced healthcare materials. 4: 1600–27. doi:10.1002/adhm.201500272. PMID 26010739.

- Topmiller, Jennifer L.; Dunn, Kevin H. (9 December 2013). "Controlling Exposures to Workers Who Make or Use Nanomaterials". National Institute of Occupational Safety and Health. Retrieved 6 January 2015.

- "Current Strategies for Engineering Controls in Nanomaterial Production and Downstream Handling Processes" (PDF). National Institute of Occupational Safety and Health. November 2013. Retrieved 6 January 2015.

Diverse Applications of Ceramic Engineering

The diverse applications of ceramic engineering are bulletproof vest, medical implant, artificial bone, insulator and airframe. Ceramic engineering is an emerging field of study; the following chapter will not only provide an overview, it will also delve deep into the applications related to it.

Bulletproof Vest

A ballistic vest or bullet-resistant vest, often called a bulletproof vest, is an item of personal armor that helps absorb the impact and reduce or stop penetration to the body from firearm-fired projectiles- and shrapnel from explosions, and is worn on the torso. Soft vests are made of many layers of woven or laminated fibers and can protect the wearer from small-caliber handgun and shotgun projectiles, and small fragments from explosives such as hand grenades.

The Improved Outer Tactical Vest (IOTV), here in Universal Camouflage Pattern, is issued to U.S. Army soldiers

These vests often have a ballistic plate inserted into the vest. Metal or ceramic plates can be used with a soft vest, providing additional protection against rifle rounds, and metallic components or tightly woven fiber layers can give soft armour resistance to stab and slash attacks from knives and similar close-quarter weapons. Soft vests are commonly worn by police forces, private citizens who are at risk of being shot (e.g., national leaders), security guards, and bodyguards, whereas hard-plate reinforced vests are mainly worn by combat soldiers, police tactical units, and hostage rescue teams.

Body armor may combine a ballistic vest with other items of protective clothing, such as a combat helmet. Vests intended for police and military use may also include ballistic shoulder and side protection armor components, and bomb disposal officers wear heavy armor and helmets with face visors and spine protection.

Overview

Ballistic vests use layers of very strong fibers to "catch" and deform a bullet, mushrooming it into a dish shape, and spreading its force over a larger portion of the vest fiber. The vest absorbs the energy from the deforming bullet, bringing it to a stop before it can completely penetrate the textile matrix. Some layers may be penetrated but as the bullet deforms, the energy is absorbed by a larger and larger fiber area.

While a vest can prevent bullet penetration, the vest and wearer still absorb the bullet's energy. Even without penetration, modern pistol bullets contain enough energy to cause blunt force trauma under the impact point. Vest specifications will typically include both penetration resistance requirements and limits on the amount of impact energy that is delivered to the body.

Vests designed for bullets offer little protection against blows from sharp implements, such as knives, arrows or ice picks, or from bullets manufactured of non-deformable materials, e.g., those containing a steel core instead of lead. This is because the impact force of these objects stays concentrated in a relatively small area, allowing them to puncture the fiber layers of most bullet-resistant fabrics. By contrast, stab vests provide better protection against sharp implements, but are generally less effective against bullets.

Textile vests may be augmented with metal (steel or titanium), ceramic or polyethylene plates that provide extra protection to vital areas. These hard armor plates have proven effective against all handgun bullets and a range of rifles. These upgraded ballistic vests have become standard in military use, as soft body armor vests are ineffective against military rifle rounds. Prison guards and police often wear vests which are designed specifically against bladed weapons and sharp objects. These vests may incorporate coated and laminated para-aramid textiles or metallic components.

History

Early Modern Era

In 1538, Francesco Maria della Rovere commissioned Filippo Negroli to create a bulletproof vest. In 1561, Maximilian II, Holy Roman Emperor is recorded as testing his armor against gun-fire. Similarly, in 1590 Sir Henry Lee expected his Greenwich armor to be "pistol proof". Its actual effectiveness was controversial at the time. The etymology of "bullet" and the adjective form of "proof" in the late 16th century would suggest that the term "bulletproof" originated shortly thereafter.

During the English Civil War Oliver Cromwell's Ironside cavalry were equipped with Capeline helmets and musket-proof cuirasses which consisted of two layers of armor plate (in later studies involving X-ray a third layer was discovered which was placed in between the outer and inner layer). The outer layer was designed to absorb the bullet's energy and the thicker inner layer stopped further penetration. The armor would be left badly dented but still serviceable. One of the first recorded descriptions of soft armor use was found in medieval Japan, with the armor having been manufactured from silk.

Industrial Era

One of the first commercially sold bulletproof armour was produced by a tailor in Dublin, Ireland in the 1840s. *The Cork Examiner* reported on his line of business in December 1847:

> The daily melancholy announcements of assassination that are now disgracing the country, and the murderers permitted to walk quietly away and defy the law, have induced me to get constructed a garment, shot and ball proof, so that every man can be protected, and enabled to return the fire of the assassin, and thus soon put a stop to the cowardly conduct which has deprived society of so many excellent and valuable lives, spreading terror and desolation through the country. I hope in a few days to have a specimen garment on view at my warerooms.

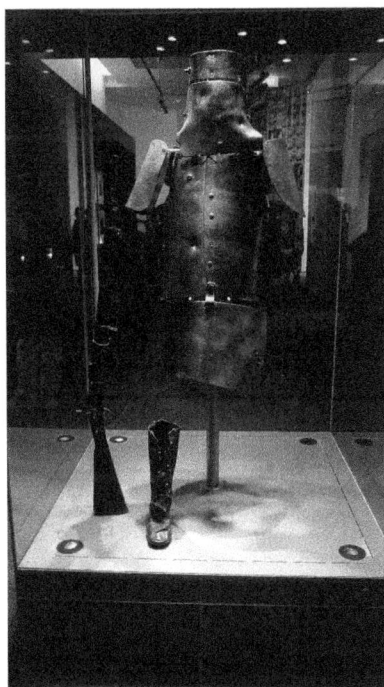

Ned Kelly's Ploughboard Ballistic Suit

Another soft ballistic vest, Myeonje baegab, was invented in Joseon, Korea in the 1860s shortly after the French campaign against Korea. The Heungseon Daewongun ordered development of bullet-proof armor because of increasing threats from Western armies. Kim Gi-Doo and Gang Yoon found that cotton could protect against bullets if 10 layers of cotton fabric were used. The vests were used in battle during the United States expedition to Korea, when the US Navy attacked Ganghwa Island in 1871. The US Navy captured one of the vests and took it to the US, where it was stored at the Smithsonian Museum until 2007. The vest has since been sent back to Korea and is currently on display to the public.

Simple ballistic armor was sometimes constructed by criminals. During the 1880s, a gang of Australian bushrangers led by Ned Kelly made basic armour from plough blades. By this time the Victorian Government had a reward for the capture of a member of the Kelly Gang at £8,000 (equivalent to $2 million Australian dollars in 2005). One of the stated aims of Kelly was the establishment of a Republic in North East Victoria. Each of the four Kelly gang members had fought a siege at a hotel clad in suits of armour made from the mouldboards of ploughs. The maker's stamp (Lennon Number 2 Type) was found inside several of the plates. The men used the armour to cover their torsos, upper arms, and upper legs, and was worn with a helmet. The suits were roughly made on a creek bed using a makeshift forge and a stringy-bark log as a muffled anvil. They had a mass of around 44 kg (96 lb), making the wearer a spectacular sight yet proved too unwieldy during a police raid at Glenrowan. Their armour deflected many hits with none penetrating, but eventually was of no use as the suits lacked protection for the legs and hands. American outlaw and gunfighter Jim Miller, was infamous for wearing a steel breastplate over his frock coat as a form of body armor. This plate saved Miller on two occasions; and it proved to be highly resistant to pistol bullets and shotguns. One example can be seen in his gun battle with a sheriff named George A. "Bud" Frazer, where the plate managed to deflect all bullets from the lawmen's six shooter after he tried to empty on Miller's chest.

Test of a 1901 vest designed by Jan Szczepanik, in which a 7 mm revolver is fired at a person wearing the vest

In 1881, Tombstone physician George E. Goodfellow noticed that a faro dealer Charlie Storms who was shot twice by Luke Short had one bullet stopped by a silk handkerchief in his breast pocket that prevented that bullet from penetrating. In 1887, he wrote an article titled *Impenetrability of Silk to Bullets* for the *Southern California Practitioner* documenting the first known instance of bulletproof fabric. He experimented with silk vests resembling medieval gambesons, which used 18 to 30 layers of silk fabric to protect the wearers from penetration.

Fr. Kazimierz Żegleń used Goodfellow's findings to develop a bulletproof vest made of silk fabric at the end of the 19th century, which could stop the relatively slow rounds from black powder handguns. The vests cost $800 USD each in 1914, a small fortune given the $20/1oz-Au exchange-rate back then, equivalent to ~$50,000 circa 2016, exceeding mean annual income. On 28 June 1914, Archduke Franz Ferdinand of Austria, heir to the throne of Austria-Hungary was fatally shot; despite owning a silk bulletproof vest, which tests by Britain's Royal Armouries indicate would likely have stopped a bullet of that era, and despite being aware of potential threats to his life including an attempted assassination of his uncle a few years earlier, Ferdinand was not wearing his on that fateful day.

A similar vest, made by Polish inventor Jan Szczepanik in 1901, saved the life of Alfonso XIII of Spain when he was shot by an attacker. By 1900, gangsters were wearing $800 silk vests to protect themselves.

First World War

World War I German Infanterie-Panzer, 1918

The combatants of World War I started the war without any attempt at providing the soldiers with body armor. Various private companies advertised body protection suits such as the Birmingham Chemico Body Shield, although these products were generally far too expensive for an average soldier.

The first official attempts at commissioning body armor were made in 1915 by the British Army Design Committee, in particular a 'Bomber's Shield' for the use of bomber pilots who were notoriously under-protected in the air from anti-aircraft bullets and shrapnel. The Experimental Ordnance Board also reviewed potential materials for bullet and fragment proof armor, such as steel plate. A 'necklet' was successfully issued on a small scale (due to cost considerations), which protected the neck and shoulders from bullets traveling at 600 feet per second with interwoven layers of silk and cotton stiffened with resin. The Dayfield body shield entered service in 1916 and a hardened breastplate was introduced the following year.

The British army medical services calculated towards the end of the War, that three quarters of all battle injuries could have been prevented if an effective armor had been issued.

The French also experimented with steel visors attached to the Adrian helmet and 'abdominal armor' designed by General Adrian. These failed to be practical, because they severely impeded the soldier's mobility. The Germans officially issued body armor in the shape of nickel and silicon armor plates that was called 'Lobster armor' from late 1916. These were similarly too heavy to be practical for the rank-and-file, but were used by static units, such as sentries and occasionally the machine-gunners. An improved version, the Infantrie-Panzer, was introduced in 1918, with hooks for equipment.

The United States developed several types of body armor, including the chrome nickel steel Brewster Body Shield, which consisted of a breastplate and a headpiece and could withstand Lewis Gun bullets at 2,700 ft/s (820 m/s), but was clumsy and heavy at 40 lb (18 kg). A scaled waistcoat of overlapping steel scales fixed to a leather lining was also designed; this armor weighed 11 lb (5.0 kg), fit close to the body, and was considered more comfortable.

Testing a bulletproof vest in Washington, D.C. September 1923.

During the late 1920s through the early 1930s, gunmen from criminal gangs in the United States began wearing less-expensive vests made from thick layers of cotton padding and cloth. These early vests could absorb the impact of handgun rounds such as .22 Long Rifle, .25 ACP, .32 S&W Long, .32 S&W, .380 ACP, .38 Special and .45 ACP traveling at speeds of up to 300 m/s (980 ft/s). To overcome these vests, law enforcement agents such as the FBI began using the newer and more powerful .38 Super, and later the .357 Magnum cartridge.

Second World War

A Japanese vest, which used overlapping armour plates

In 1940, the Medical Research Council in Britain proposed the use of a lightweight suit of armor for general use by infantry, and a heavier suit for troops in more dangerous positions, such as anti-aircraft and naval gun crews. By February 1941, trials had begun on body armor made of manganese steel plates. Two plates covered the front area and one plate on the lower back protected the kidneys and other vital organs. Five thousand sets were made and evaluated to almost unanimous approval - as well as providing adequate protection, the armor didn't severely impede the mobility of the soldier and were reasonably comfortable to wear. The armor was introduced in 1942 although the demand for it was later scaled down. The Canadian Army in northwestern Europe also adopted this armor for the medical personnel of the 2nd Canadian Infantry Division.

The British company Wilkinson Sword began to produce flak jackets for bomber crew in 1943 under contract with the Royal Air Force. It was realised that the majority of pilot deaths in the air was due to low velocity fragments rather than bullets. Surgeon of the United States Army Air Forces, Colonel M. C. Grow, stationed in Britain, thought that many wounds he was treating could have been prevented by some kind of light armor. Two types of armor were issued for different specifications. These jackets were made of nylon fabric and capable of stopping flak and shrapnel, but were not designed to stop bullets. Although they were considered too bulky for pilots using the Avro Lancaster bombers, they were adopted by United States Army Air Forces.

In the early stages of World War II, the United States also designed body armor for infantrymen, but most models were too heavy and mobility-restricting to be useful in the field and incompatible with existing required equipment. Near the middle of 1944, development of infantry body armor in the United States restarted. Several vests were produced for the US military, including but not limited to the T34, the T39, the T62E1, and the M12. The United States developed a vest using Doron Plate, a fiberglass-based laminate. These vests were first used in the Battle of Okinawa in 1945.

Sn-42 Body Armor

The Soviet Armed Forces used several types of body armor, including the SN-42 ("Stalnoi Nagrudnik" is Russian for "steel breastplate", and the number denotes the design year). All were tested, but only the SN-42 was put in production. It consisted of two pressed steel plates that protected the front torso and groin. The plates were 2 mm thick and weighed 3.5 kg (7.7 lb). This armor was supplied to SHISBr (assault engineers). The SN armor protected wearers from 9×19mm bullets fired by an MP 40 at around 100 meters, which made it useful in urban battles such as the Battle of Stalingrad. However, the SN's weight made it impractical for infantry in the open.

Postwar

During the Korean War several new vests were produced for the United States military, including the M-1951, which made use of fibre-reinforced plastic or aluminium segments woven into a nylon vest. These vests represented "a vast improvement on weight, but the armor failed to stop bullets and fragments very successfully," although officially they were claimed to be able to stop 7.62×25mm Tokarev pistol rounds at the muzzle. Developed by Natick Laboratories and introduced in 1967, T65-2 plate carriers were the first vests designed to hold hard ceramic plates, making them capable of stopping 7 mm rifle rounds.

Two American GIs wearing M1951 bullet-proof vests on Triangle Hill in Korea.

These "Chicken Plates" were made of either boron carbide, silicon carbide, or aluminium oxide. They were issued to the crew of low-flying aircraft, such as the UH-1 and UC-123, during the Vietnam War.

In 1969, American Body Armor was founded and began to produce a patented combination of quilted nylon faced with multiple steel plates. This armor configuration was marketed to American law enforcement agencies by Smith & Wesson under the trade name "Barrier Vest." The Barrier Vest was the first police vest to gain wide use during high threat police operations.

In 1971, research chemist Stephanie Kwolek discovered a liquid crystalline polymer solution. Its exceptional strength and stiffness led to the invention of Kevlar®, a synthetic fibre, woven into a fabric and layered, that, by weight, has five times the tensile strength of steel. In the mid-1970s, DuPont the company which employed Kwolek introduced Kevlar. Immediately Kevlar was incorporated into a National Institute of Justice (NIJ) evaluation program to provide lightweight, able body armor to a test pool of American law enforcement officers to ascertain if everyday able wearing was possible. Lester Shubin, a program manager at the NIJ, managed this law enforcement feasibility study within a few selected large police agencies, and quickly determined that Kevlar body armor could be comfortably worn by police daily, and would save lives.

In 1975 Richard A. Armellino, the founder of American Body Armor, marketed an all Kevlar vest called the K-15, consisting of 15 layers of Kevlar that also included a 5" × 8" ballistic steel "Shok Plate" positioned vertically over the heart and was issued US Patent #3,971,072 for this innovation. Similarly sized and positioned "trauma plates" are still used today on the front ballistic panels of most able vests, reducing blunt trauma and increasing ballistic protection in the center-mass heart/sternum area.

In 1976, Richard Davis, founder of Second Chance Body Armor, designed the company's first all-Kevlar vest, the Model Y. The lightweight, able vest industry was launched and a new form of daily protection for the modern police officer was quickly adapted. By the mid-to-late 1980s, an estimated 1/3 to 1/2 of police patrol officers wore able vests daily. By 2006, more than 2,000 documented police vest "saves" were recorded, validating the success and efficiency of lightweight able body armor as a standard piece of everyday police equipment.

Recent years

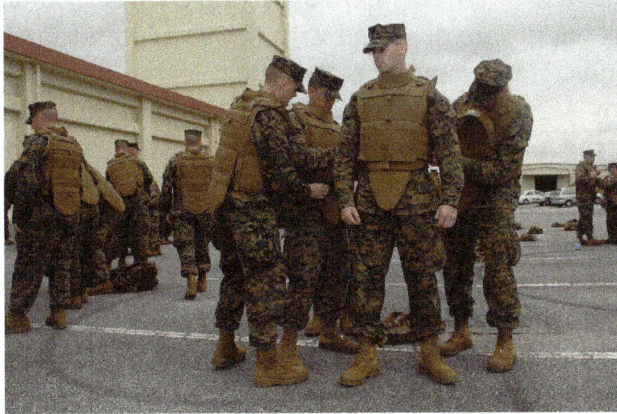

US Marines being issued the MTV at Camp Foster, Okinawa

During the 1980s, the US military issued the PASGT kevlar vest, rated at NIJ level IIA,being able to stop pistol rounds (including 9 mm FMJ) and fragmentation. West Germany issued a similar rated vest called the Splitterschutzweste.

Kevlar soft armor had its shortcomings because if "large fragments or high velocity bullets hit the vest, the energy could cause life-threatening, blunt trauma injuries" in selected, vital areas. Ranger Body Armor was developed for the American military in 1991. Although it was the second modern US body armor that was able to stop rifle caliber rounds and still be light enough to be worn by infantry soldiers in the field, it still had its flaws: "it was still heavier than the concurrently issued PASGT (Personal Armor System for Ground Troops) anti-fragmentation armor worn by regular infantry and … did not have the same degree of ballistic protection around the neck and shoulders." The format of Ranger Body Armor (and more recent body armor issued to US special operations units) highlights the trade-offs between force protection and mobility that modern body armor forces organizations to address.

A U.S. Army working dog, wearing a bullet-resistant vest, clears a building in Afghanistan.

Newer armor issued by the United States armed forces to large numbers of troops includes the United States Army's Improved Outer Tactical Vest and the United States Marine Corps Modular Tactical Vest. All of these systems are designed with the vest intended to provide protection from fragments and pistol rounds. Hard ceramic plates, such as the Small Arms Protective Insert, as used with Interceptor Body Armor, are worn to protect the vital organs from higher level threats. These threats mostly take the form of high velocity and armor-piercing rifle rounds. Similar types of protective equipment have been adopted by modern armed forces over the world.

Since the 1970s, several new fibers and construction methods for bulletproof fabric have been developed besides woven Kevlar, such as DSM's Dyneema, Honeywell's Gold Flex and Spectra, Teijin Twaron's Twaron, Pinnacle Armor's Dragon Skin, and Toyobo's Zylon. These newer materials are advertised as being lighter, thinner and more resistant than Kevlar, although they are much more expensive. The US military has developed body armor for the working dogs who aid GIs in battle.

Since 2004, U.S. Special Operations Command has been at work on a new full-body armor that will rely on rheology, or the technology behind the elasticity of liquids in skin care and automotive products. Named TALOS, this new technology may be used in the future.

Performance Standards

Indonesian Special Police "Brimob" personnel and an officer (left) with bulletproof vests in Jakarta during the 2016 Jakarta attacks

Due to the various types of projectile, it is often inaccurate to refer to a particular product as "bulletproof" because this implies that it will protect against any and all threats. Instead, the term bullet resistant is generally preferred.

Body armor standards are regional. Around the world ammunition varies and as a result the armor testing must reflect the threats found locally. Law enforcement statistics show that many shootings where officers are injured or killed involve the officer's own weapon. As a result, each law enforcement agency or para-military organization will have their own standard for armor performance if only to ensure that their armor protects them from their own weapons. While many standards exist, a few standards are widely used as models. The US National Institute of Justice ballistic and stab documents are examples of broadly accepted standards. In addition to the NIJ, the UK Home Office Scientific Development Branch (HOSDB – formerly the Police Scientific Development Branch (PSDB)) standards are used by a number of other countries and organizations. These "model" standards are usually adapted by other counties by incorporation of the basic test methodologies with modification of the bullets that are required for test. NIJ Standard-0101.06 has specific performance standards for bullet resistant vests used by law enforcement. This rates vests on the following scale against penetration and also blunt trauma protection (deformation):

Armor Level	Protection
Type I .22 LR .380 ACP	This armor would protect against • 2.6 g (40 gr) .22 Long Rifle Lead Round Nose (LR LRN) bullets at a velocity of 329 m/s (1080 ft/s ± 30 ft/s) • 6.2 g (95 gr) .380 ACP Full Metal Jacketed Round Nose (FMJ RN) bullets at a velocity of 322 m/s (1055 ft/s ± 30 ft/s). It is no longer part of the standard.
Type IIA 9×19mm .40 S&W .45 ACP	New armor protects against: • 8 g (124 gr) 9×19mm Parabellum Full Metal Jacketed Round Nose (FMJ RN) bullets at a velocity of 373 m/s ± 9.1 m/s (1225 ft/s ± 30 ft/s) • 11.7 g (180 gr) .40 S&W Full Metal Jacketed (FMJ) bullets at a velocity of 352 m/s ± 9.1 m/s (1155 ft/s ± 30 ft/s) • 14.9 g (230 gr) .45 ACP Full Metal Jacketed (FMJ) bullets at a velocity of 275 m/s ± 9.1 m/s (900 ft/s ± 30 ft/s). Conditioned armor protects against • 8 g (124 gr) 9 mm FMJ RN bullets at a velocity of 355 m/s ± 9.1 m/s (1165 ft/s ± 30 ft/s) • 11.7 g (180 gr) .40 S&W FMJ bullets at a velocity of 325 m/s ± 9.1 m/s (1065 ft/s ± 30 ft/s) • 14.9 g (230 gr) .45 ACP Full Metal Jacketed (FMJ) bullets at a velocity of 259 m/s ± 9.1 m/s (850 ft/s ± 30 ft/s). It also provides protection against the threats mentioned in [Type I].
Type II 9mm +P . 3 5 7 Magnum	New armor protects against • 8 g (124 gr) 9 mm FMJ RN bullets at a velocity of 398 m/s ± 9.1 m/s (1305 ft/s ± 30 ft/s) • 10.2 g (158 gr) .357 Magnum Jacketed Soft Point bullets at a velocity of 436 m/s ± 9.1 m/s (1430 ft/s ± 30 ft/s). Conditioned armor protects against • 8 g (124 gr) 9 mm FMJ RN bullets at a velocity of 379 m/s ±9.1 m/s (1245 ft/s ± 30 ft/s) • 10.2 g (158 gr) .357 Magnum Jacketed Soft Point bullets at a velocity of 408 m/s ±9.1 m/s (1340 ft/s ± 30 ft/s). It also provides protection against the threats mentioned in [Types I and IIA].

Type IIIA	New armor protects against
.357 SIG . 4 4 Magnum 1 0 m m Auto 7H21	• 8.1 g (125 gr) .357 SIG FMJ Flat Nose (FN) bullets at a velocity of 448 m/s ± 9.1 m/s (1470 ft/s ± 30 ft/s) • 15.6 g (240 gr) .44 Magnum Semi Jacketed Hollow Point (SJHP) bullets at a velocity of 436 m/s (1430 ft/s ± 30 ft/s). Conditioned armor protects against • 8.1 g (125 gr) .357 SIG FMJ Flat Nose (FN) bullets at a velocity of 430 m/s ± 9.1 m/s (1410 ft/s ± 30 ft/s) • 15.6 g (240 gr) .44 Magnum Semi Jacketed Hollow Point (SJHP) bullets at a velocity of 408 m/s ± 9.1 m/s (1340 ft/s ± 30 ft/s). It also provides protection against most handgun threats, as well as the threats mentioned in [Types I, IIA, and II].
Type III Rifles 7H31 L e h i g h .45-70	Conditioned armor protects against • 8.0 g (123 gr) 7.62×39mm (the ubiquitous AK-47 round) FMJ at a velocity of 738 m/s (2,421 ft/s. • 9.6 g (148 gr) 7.62×51mm NATO M80 ball bullets at a velocity of 847 m/s ± 9.1 m/s (2780 ft/s ± 30 ft/s). • 19.7 g (305 gr) .45-70 solid copper bullets at velocity of 610 m/s (2000 ft/s). It also provides protection against the threats mentioned in [Types I, IIA, II, and IIIA].
Type IV A r m o r Piercing Rifle	Conditioned armor protects against • 10.8 g (166 gr) .30-06 Springfield M2 armor-piercing (AP) bullets at a velocity of 878 m/s ± 9.1 m/s (2880 ft/s ± 30 ft/s). It also provides at least single hit protection against the threats mentioned in [Types I, IIA, II, IIIA, and III].

NIJ standards are used for law enforcement armors. The US and NATO military armor designs are tested using a standard set of test methods under ARMY MIL-STD-662F and STANAG 2920 Ed2. This approach defines the test process under the 662F/2920 standard. Each armor program can select a unique series of projectiles and velocities as required. The DOD and MOD armor programs-of-record (MTV for example) procure armor using these test standards. In addition, special requirements can be defined under this process for armors for flexible rifle protection, fragment protection for the extremities, etc. These military procurement requirements do not relate to NIJ, HOSDB or ISO law enforcement armor standards, test methods, garment size, projectiles or velocities.

In addition to the NIJ and HOSDB law enforcement armor standards, other important standards include German Police TR-Technische Richtlinie, Draft ISO prEN ISO 14876, and Underwriters Laboratories (UL Standard 752).

Textile armor is tested for both penetration resistance by bullets and for the impact energy transmitted to the wearer. The "backface signature," or transmitted impact energy, is measured by shooting armor mounted in front of a backing material, typically oil-based modeling clay. The clay is used at a controlled temperature and verified for impact flow before testing. After the armor is impacted with the test bullet, the vest is removed from the clay and the depth of the indentation in the clay is measured.

The backface signature allowed by different test standards can be difficult to compare. Both the clay materials and the bullets used for the test are not common. In general the British, German and other European standards allow 20–25 mm of backface signature, while the US-NIJ standards allow for 44 mm, which can potentially cause internal injury. The allowable backface signature for body armor has been controversial from its introduction in the first NIJ test standard and the debate as to the relative importance of penetration-resistance vs. backface signature continues in the medical and testing communities.

In general a vest's textile material temporarily degrades when wet. Neutral water at room temp does not affect para-aramid or UHMWPE but acidic, basic and some other solutions can permanently reduce para-aramid fiber tensile strength. (As a result of this, the major test standards call for wet testing of textile armor.) Mechanisms for this wet loss of performance are not known. Vests that will be tested after ISO type water immersion tend to have heat sealed enclosures and those that are tested under NIJ type water spray methods tend to have water resistant enclosures.

From 2003 to 2005, a large study of the environmental degradation of Zylon armor was undertaken by the US-NIJ. This concluded that water, long-term use, and temperature exposure significantly affect tensile strength and the ballistic performance of PBO or Zylon fiber. This NIJ study on vests returned from the field demonstrated that environmental effects on Zylon resulted in ballistic failures under standard test conditions.

Ballistic Testing V50 and V0

Measuring the ballistic performance of armor is based on determining the kinetic energy of a bullet at impact ($E_k = \frac{1}{2}mv^2$). Because the energy of a bullet is a key factor in its penetrating capacity, velocity is used as the primary independent variable in ballistic testing. For most users the key measurement is the velocity at which no bullets will penetrate the armor. Measuring this zero penetration velocity (v_0) must take into account variability in armor performance and test variability. Ballistic testing has a number of sources of variability: the armor, test backing materials, bullet, casing, powder, primer and the gun barrel, to name a few.

Variability reduces the predictive power of a determination of V0. If for example, the v_0 of an armor design is measured to be 1,600 ft/s (490 m/s) with a 9 mm FMJ bullet based on 30 shots, the test is only an estimate of the real v_0 of this armor. The problem is variability. If the v_0 is tested again with a second group of 30 shots on the same vest design, the result will not be identical.

Only a single low velocity penetrating shot is required to reduce the v_0 value. The more shots made the lower the v_0 will go. In terms of statistics, the zero penetration velocity is the tail end of the distribution curve. If the variability is known and the standard deviation can be calculated, one can rigorously set the V0 at a confidence interval. Test Standards now define how many shots must be used to estimate a v_0 for the armor certification. This procedure defines a confidence interval of an estimate of v_0.

v_0 is difficult to measure, so a second concept has been developed in ballistic testing called the ballistic limit (v_{50}). This is the velocity at which 50 percent of the shots go through and 50 percent are stopped by the armor. US military standard MIL-STD-662F V50 Ballistic Test define a commonly used procedure for this measurement. The goal is to get three shots that penetrate that are slower than a second faster group of three shots that are stopped by the armor. These three high stops and three low penetrations can then be used to calculate a v_{50} velocity.

In practice this measurement of v_{50} requires 1–2 vest panels and 10–20 shots. A very useful concept in armor testing is the offset velocity between the v_0 and v_{50}. If this offset has been measured for an armor design, then v_{50} data can be used to measure and estimate changes in v_0. For vest manufacturing, field evaluation and life testing both v_0 and v_{50} are used. However, as a result of the simplicity of making v_{50} measurements, this method is more important for control of armor after certification.

Military Testing: Fragment Ballistics

After the Vietnam War, military planners developed a concept of "Casualty Reduction". The large body of casualty data made clear that in a combat situation, fragments, not bullets, were the most important threat to soldiers. After WWII, vests were being developed and fragment testing was in its early stages. Artillery shells, mortar shells, aerial bombs, grenades, and antipersonnel mines are all fragmentation devices. They all contain a steel casing that is designed to burst into small steel fragments or shrapnel, when their explosive core detonates. After considerable effort measuring fragment size distribution from various NATO and Soviet bloc munitions, a fragment test was developed. Fragment simulators were designed, and the most common shape is a right circular cylinder or RCC simulator. This shape has a length equal to its diameter. These RCC Fragment Simulation Projectiles (FSPs) are tested as a group. The test series most often includes 2 grain (0.13 g), 4 grain (0.263 g), 16 grain (1.0 g), and 64 grain (4.2 g) mass RCC FSP testing. The 2-4-16-64 series is based on the measured fragment size distributions.

German Policemen in bulletproof vests on guard duty at a military hospital.

The second part of "Casualty Reduction" strategy is a study of velocity distributions of fragments from munitions. Warhead explosives have blast speeds of 20,000 ft/s (6,100 m/s) to 30,000 ft/s (9,100 m/s). As a result, they are capable of ejecting fragments at very high speeds of over 3,300 ft/s (1,000 m/s), implying very high energy (where the energy of a fragment is ½ mass × velocity², neglecting rotational energy). The military engineering data showed that, like the fragment size, the fragment velocities had characteristic distributions. It is possible to segment the fragment output from a warhead into velocity groups. For example, 95% of all fragments from a bomb blast under

4 grains (0.26 g) have a velocity of 3,000 ft/s (910 m/s) or less. This established a set of goals for military ballistic vest design.

The random nature of fragmentation required the military vest specification to trade off mass vs. ballistic-benefit. Hard vehicle armor is capable of stopping all fragments, but military personnel can only carry a limited amount of gear and equipment, so the weight of the vest is a limiting factor in vest fragment protection. The 2-4-16-64 grain series at limited velocity can be stopped by an all-textile vest of approximately 5.4 kg/m² (1.1 lb/ft²). In contrast to the design of vest for deformable lead bullets, fragments do not change shape; they are steel and can not be deformed by textile materials. The 2-grain (0.13 g) FSP (the smallest fragment projectile commonly used in testing) is about the size of a grain of rice; such small fast moving fragments can potentially slip through the vest, moving between yarns. As a result, fabrics optimized for fragment protection are tightly woven, although these fabrics are not as effective at stopping lead bullets.

Backing Materials for Testing

Ballistic

One of the critical requirements in soft ballistic testing is measurement of "back side signature" (i.e. energy delivered to tissue by a non-penetrating projectile) in a deformable backing material placed behind the targeted vest. The majority of military and law enforcement standards have settled on an oil/clay mixture for the backing material, known as Roma Plastilena. Although harder and less deformable than human tissue, Roma represents a "worst case" backing material when plastic deformations in the oil/clay are low (less than 20 mm). (Armor placed over a harder surface is more easily penetrated.) The oil/clay mixture of "Roma" is roughly twice the density of human tissue and therefore does not match its specific gravity, however "Roma" is a plastic material that will not recover its shape elastically, which is important for accurately measuring potential trauma through back side signature.

The selection of test backing is significant because in flexible armor, the body tissue of a wearer plays an integral part in absorbing the high energy impact of ballistic and stab events. However the human torso has a very complex mechanical behavior. Away from the rib cage and spine, the soft tissue behavior is soft and compliant. In the tissue over the sternum bone region, the compliance of the torso is significantly lower. This complexity requires very elaborate bio-morphic backing material systems for accurate ballistic and stab armor testing. A number of materials have been used to simulate human tissue in addition to Roma. In all cases, these materials are placed behind the armor during test impacts and are designed to simulate various aspects of human tissue impact behavior.

One important factor in test backing for armor is its hardness. Armor is more easily penetrated in testing when backed by harder materials, and therefore harder materials, such as Roma clay, represent more conservative test methods.

Backer type	Materials	Elastic/ plastic	Test type	Specific gravity	Relative hardness vs gelatin	Application
Roma Plastilina Clay #1	Oil/Clay mixture	Plastic	Ballistic and Stab	>2	Moderately hard	Back face signature measurement. Used for most standard testing

10% gelatin	Animal protein gel	Visco-elastic	Ballistic	~1 (90% water)	Softer than baseline	Good simulant for human tissue, hard to use, expensive. Required for FBI test methods
20% gelatin	Animal protein gel	Visco-elastic	Ballistic	~1 (80% water)	Baseline	Good simulant for skeletal muscle. Provides dynamic view of event.
HOSDB-NIJ Foam	Neoprene foam, EVA foam, sheet rubber	Elastic	Stab	~1	Slightly harder than gelatin	Moderate agreement with tissue, easy to use, low in cost. Used in stab testing
Silicone gel	Long chain silicone polymer	Visco-elastic	Biomedical	~1.2	Similar to gelatin	Biomedical testing for blunt force testing, very good tissue match
Pig or Sheep animal testing	Live tissue	Various	Research	~1	Real tissue is variable	Very complex, requires ethical review for approval

Stab

Stab and spike armor standards have been developed using 3 different backing materials. The Draft EU norm calls out Roma clay, The California DOC called out 60% ballistic gelatin and the current standard for NIJ and HOSDB calls out a multi-part foam and rubber backing material.

- Using Roma clay backing, only metallic stab solutions met the 109 joule Calif. DOC ice pick requirement

- Using 10% Gelatin backing, all fabric stab solutions were able to meet the 109 joule Calif. DOC ice pick requirement.

- Most recently the Draft ISO prEN ISO 14876 norm selected Roma as the backing for both ballistics and stab testing.

This history helps explain an important factor in Ballistics and Stab armor testing, backing stiffness affects armor penetration resistance. The energy dissipation of the armor-tissue system is Energy = Force × Displacement when testing on backings that are softer and more deformable the total impact energy is absorbed at lower force. When the force is reduced by a softer more compliant backing the armor is less likely to be penetrated. The use of harder Roma materials in the ISO draft norm makes this the most rigorous of the stab standards in use today.

Rifle Resistant Armor

Because of the limitations of the technology a distinction is made between handgun protection and rifle protection. Broadly rifle resistant armor is of three basic types: ceramic plate-based systems, steel plate with spall fragmentation protective coating, and hard fiber-based laminate systems. Many rifle armor components contain both hard ceramic components and laminated textile materials used together. Various ceramic materials types are in use, however: aluminum oxide, boron carbide and silicon carbide are the most common. The fibers used in these systems are the same as found in soft textile armor. However, for rifle protection high pressure lamination of ultra high molecular weight polyethylene with a Kraton matrix is the most common.

The Small Arms Protective Insert (SAPI) and the enhanced SAPI plate for the US DOD generally has this form. Because of the use of ceramic plates for rifle protection, these vests are 5–8 times as heavy on an area basis as handgun protection. The weight and stiffness of rifle armor is a major technical challenge. The density, hardness and impact toughness are among the materials properties that are balanced to design these systems. While ceramic materials have some outstanding properties for ballistics they have poor fracture toughness. Failure of ceramic plates by cracking must also be controlled. For this reason many ceramic rifle plates are a composite. The strike face is ceramic with the backface formed of laminated fiber and resin materials. The hardness of the ceramic prevents the penetration of the bullet while the tensile strength of the fiber backing helps prevent tensile failure. Examples of rifle resistant outer vests include the Interceptor body armor and the Improved Outer Tactical Vest.

Versus Armor-piercing Ammunition

The standards for armor-piercing rifle bullets aren't clear-cut, because the penetration of a bullet depends on the hardness of the target armor. However, there are a few general rules. For example, bullets with a soft lead-core and copper jacket are too easily deformed to penetrate hard materials, whereas rifle bullets manufactured with very hard core materials, like tungsten carbide, are designed for maximum penetration into hard armor. Most other core materials would have effects between lead and tungsten carbide. Many common bullets, such as the 7.62×39mm M43 standard cartridge for the AK-47 rifle, have a steel core with hardness rating ranging from Rc35 mild steel up to Rc45 medium hard steel.

Additionally, as the hardness of the bullet core increases, so must the amount of ceramic plating used to stop penetration. Like in soft ballistics, a minimum ceramic material hardness of the bullet core is required to damage their respective hard core materials, however in armor-piercing rounds the bullet core is eroded rather than deformed.

The US Department of Defense uses two classes of protection from armor-piercing rifle bullets. The first, the Small Arms Protective Insert (SAPI), called for ceramic composite plates with a mass of 20–30 kg/m² (4–5 lb/ft²). Later, the Enhanced SAPI (ESAPI) specification was developed to protect from more penetrative ammunition. ESAPI ceramic plates have a density of 35–45 kg/m² (7–9 lb/ft²), and are designed to stop bullets like the .30-06 AP (M2) with an engineered hard core.

Cercom, now BAE systems, CoorsTek, Ceradyne, Tencate, Honeywell, DSM, Pinnacle Armor and a number of other engineering companies develop and manufacture the materials for composite ceramic rifle armor.

Explosive Protection

Bomb suit being used in a training exercise

Bomb disposal officers often wear heavy armor designed to protect against most effects of a moderate sized explosion, such as bombs encountered in terror threats. Full head helmet, covering the face and some degree of protection for limbs is mandatory in addition to very strong armor for the torso. An insert to protect the spine is usually applied to the back, in case an explosion blasts the wearer. Visibility and mobility of the wearer is severely limited, as is the time that can be spent working on the device. Armor designed primarily to counter explosives is often somewhat less effective against bullets than armor designed for that purpose. The sheer mass of most bomb disposal armor usually provides some protection, and bullet-specific trauma plates are compatible with some bomb disposal suits. Bomb disposal technicians try to accomplish their task if possible using remote methods (e.g., robots, line and pulleys). Actually laying hands on a bomb is only done in an extremely life-threatening situation, where the hazards to people and critical structures cannot be lessened by using wheeled robots or other techniques.

Early "Ice Pick" Test

In the mid-1980s the state of California Department of Corrections issued a requirement for a body armor using a commercial ice pick as the test penetrator. The test method attempted to simulate the capacity of a human attacker to deliver impact energy with their upper body. As was later shown by the work of the former British PSDB, this test over stated the capacity of human attackers. The test used a drop mass or sabot that carried the ice pick. Using gravitational force, the height of the drop mass above the vest was proportional to the impact energy. This test specified 109 joules (81 ft·lb) of energy and a 7.3 kg (16 lb) drop mass with a drop height of 153 cm (60 in).

The ice pick has a 4 mm (0.16 in) diameter with a sharp tip with a 5.4 m/s (17 ft/s) terminal velocity in the test. The California standard did not include knife or cutting edge weapons in the test protocol. The test method used the oil/clay (Roma Plastilena) tissue simulant as a test backing. In this early phase only titanium and steel plate offerings were successful in addressing this requirement. Point Blank developed the first ice pick certified offerings for CA Department of Corrections in shaped titanium sheet metal. Vests of this type are still in service in US corrections facilities as of 2008.

Beginning in the early 1990s, an optional test method was approved by California which permitted the use of 10% ballistic gelatin as a replacement for Roma clay. The transition from hard, dense clay-based Roma to soft low-density gelatin allowed all textile solutions to meet this attack energy requirement. Soft all textile "ice pick" vests began to be adopted by California and other US states as a result of this migration in the test methods. It is important for users to understand that the smooth, round tip of the ice pick does not cut fiber on impact and this permits the use of textile based vests for this application.

The earliest of these "all" fabric vests designed to address this ice pick test was Warwick Mills's TurtleSkin ultra tightly woven para-aramid fabric with a patent filed in 1993. Shortly after the TurtleSkin work, in 1995 DuPont patented a medium density fabric that was designated as Kevlar Correctional. It should be noted that these textile materials do not have equal performance with cutting-edge threats and these certifications were only with ice pick and were not tested with knives.

HOSDB-Stab and Slash Standards

Parallel to the US development of "ice pick" vests, the British police, PSDB, was working on standards for knife-resistant body armor. Their program adopted a rigorous scientific approach and collected data on human attack capacity. Their ergonomic study suggested three levels of threat: 25, 35 and 45 joules of impact energy. In addition to impact energy attack, velocities were measured and were found to be 10–20 m/s (much faster than the California test). Two commercial knives were selected for use in this PSDB test method. In order to test at a representative velocity, an air cannon method was developed to propel the knife and sabot at the vest target using compressed air. In this first version, the PSDB '93 test also used oil/clay materials as the tissue simulant backing. The introduction of knives which cut fiber and a hard-dense test backing required stab vest manufacturers to use metallic components in their vest designs to address this more rigorous standard. The current standard HOSDB Body Armour Standards for UK Police (2007) Part 3: Knife and Spike Resistance is harmonized with the US NIJ OO15 standard, use a drop test method and use a composite foam backing as a tissue simulant. Both the HOSDB and the NIJ test now specify engineered blades, double-edged S1 and single-edge P1 as well as the spike.

In addition to the stab standards, HOSDB has developed a standard for slash resistance (2006). This standard, like the stab standards, is based on drop testing with a test knife in a mounting of controlled mass. The slash test uses the Stanley Utility knife or box cutter blades. The slash standard tests the cut resistance of the armor panel parallel to the direction of blade travel. The test equipment measures the force at the instant the blade tip produces a sustained slash through the vest. The criteria require that slash failure of the armor be greater than 80 newtons of force.

Combination Stab and Ballistic Vests

Vests that combined stab and ballistic protection were a significant innovation in the 1990s period of vest development. The starting point for this development were the ballistic-only offerings of that time using NIJ Level 2A, 2, and 3A or HOSDB HG 1 and 2, with compliant ballistic vest products being manufactured with areal densities of between 5.5 and 6 kg/m² (1.1 and 1.2 lb/ft² or 18 and 20 oz/ft²). However police forces were evaluating their "street threats" and requiring vests with both knife and ballistic protection. This multi-threat approach is common in the United Kingdom and other European countries and is less popular in the USA. Unfortunately for multi-threat users, the metallic array and chainmail systems that were necessary to defeat the test blades offered little ballistic performance. The multi-threat vests have areal densities are close to the sum of the two solutions separately. These vests have mass values in the 7.5–8.5 kg/m² (1.55–1.75 lb/ft²) range. Ref (NIJ and HOSDB certification listings). Rolls Royce Composites -Megit and Highmark produced metallic array systems to address this HOSDB standard. These designs were used extensively by the London Metropolitan Police Service and other agencies in the United Kingdom.

Standards Update US and UK

As vest manufactures and the specifying authorities worked with these standards, the UK and US Standards teams began a collaboration on test methods. A number of issues with the first versions of the tests needed to be addressed. The use of commercial knives with inconsistent sharpness and tip shape created problems with test consistency. As a result, two new "engineered blades" were designed that could be manufactured to have reproducible penetrating behavior. The tissue simulants, Roma clay and gelatin, were either unrepresentative of tissue or not practical for the test operators. A composite-foam and hard-rubber test backing was developed as an alternative to address these issues. The drop test method was selected as the baseline for the updated standard over the air cannon option. The drop mass was reduced from the "ice pick test" and a wrist-like soft linkage was engineered into the penetrator-sabot to create a more realistic test impact. These closely related standards were first issued in 2003 as HOSDB 2003 and NIJ 0015. (The Police Scientific Development Branch (PSDB) was renamed the Home Office Scientific Development Branch in 2004.

Metropolitan Police officers supervising World Cup, 2006

Stab and Spike Vests

These new standards created a focus on Level 1 at 25 joules (18 ft·lbf), Level 2 at 35 J (26 ft·lbf), Level 3 at 45 J (33 ft·lbf) protection as tested with the new engineered knives defined in these test documents. The lowest level of this requirement at 25 joules was addressed by a series of textile products of both wovens, coated wovens and laminated woven materials. All of these materials were based on Para-aramid fiber. The co-efficient of friction for ultra high molecular weigh polyethylene (UHMWPE) prevented its use in this application. The TurtleSkin DiamondCoat and Twaron SRM products addressed this requirement using a combination of Para-Aramid wovens and bonded ceramic grain. These ceramic-coated products do not have the flexibility and softness of un-coated textile materials.

For the higher levels of protection L2 and L3, the very aggressive penetration of the small, thin P1 blade has resulted in the continued use of metallic components in stab armor. In Germany, Mehler Vario Systems developed hybrid vests of woven para-aramid and chainmail, and their solution was selected by London's Metropolitan Police Service. Another German company BSST, in cooperation with Warwick Mills, has developed a system to meet the ballistic-stab requirement using Dyneema laminate and an advanced metallic-array system, TurtleSkin MFA. This system is currently implemented in the Netherlands. The trend in multi threat armor continues with requirements for needle protection in the Draft ISO prEN ISO 14876 norm. In many countries there is also an interest to combine military style explosive fragmentation protection with bullet-ballistics and stab requirements.

Vest sizing, Carriers and Encapsulation

In order for ballistic protection to be wearable the ballistic panels and hard rifle-resistant plates are fitted inside a special carrier. The carrier is the visible part of a ballistic vest. The most basic carrier includes pockets which hold the ballistic panels and straps for mounting the carrier on the user. There are two major types of carriers: military or tactical carriers that are worn over the shirt, and covert law enforcement type carriers that are worn under the shirt.

Military Carriers

Individual pieces comprising the Modular Tactical Vest worn by U.S. Marines, including SAPI plates (gray, at top left)

The military type of carrier, English police waistcoat carrier, or police tactical carrier most typically has a series of webbing, hook and loop, and snap type connectors on the front and back face. This permits the wearer to mount various gear to the carrier in many different configurations. This load carriage feature is an important part of uniform and operational design for police weapons teams and the military.

In addition to load carriage, this type of carrier may include pockets for neck protection, side plates, groin plates, and backside protection. Because this style of carrier is not close fitting, sizing in this system is straightforward for both men and women, making custom fabrication unnecessary.

Concealable Carriers

Law enforcement carriers in some countries are concealable. The carrier holds the ballistic panels close to the wearer's body and a uniform shirt is worn over the carrier. This type of carrier must be designed to conform closely to the officer's body shape. For concealable armor to conform to the body it must be correctly fitted to a particular individual. Many programs specify full custom measurement and manufacturing of armor panels and carriers to ensure good fit and comfortable armor. Officers who are either female or significantly overweight have more difficulty in getting accurately measured and having comfortable armor fabricated.

Vest Slips

A third textile layer is often found between the carrier and the ballistic components. The ballistic panels are covered in a coated pouch or slip. This slip provides the encapsulation of the ballistic materials. Slips are manufactured in two types: heat sealed hermetic slips and simple sewn slips. For some ballistic fibers such as Kevlar the slip is a critical part of the system. The slip prevents moisture from the user's body from saturating the ballistic materials. This protection from moisture cycling increases the useful life of the armor.

Research

Progress in Fiber Math

In recent years, advances in material science have opened the door to the idea of a literal "bullet-proof vest" able to stop handgun and rifle bullets with a soft textile vest, without the assistance of additional metal or ceramic plating. However, progress is moving at a slower rate compared to other technical disciplines. The most recent offering from Kevlar, Protera, was released in 1996. Current soft body armor can stop most handgun rounds (which has been the case for roughly 15 years), but armor plates are needed to stop rifle rounds and steel-core handgun rounds such as $7.62 \times 25mm$. The para-aramids have not progressed beyond the limit of 23 grams per denier in fiber tenacity.

Modest ballistic performance improvements have been made by new producers of this fiber type. Much the same can be said for the UHMWPE material; the basic fiber properties have only advanced to the $30-35$ g/d range. Improvements in this material have been seen in the development of cross-plied non-woven laminate, e.g. Spectra Shield. The major ballistic performance advance of fiber PBO is known as a "cautionary tale" in materials science. This fiber permitted the design of handgun soft armor that was $30-50\%$ lower in mass as compared to the aramid and UHMWPE materials. However this higher tenacity was delivered with a well-publicized weakness in environmental durability.

Akzo-Magellan (now DuPont) teams have been working on fiber called M5 fiber; however, its announced startup of its pilot plant has been delayed more than 2 years. Data suggests if the M5 material can be brought to market, its performance will be roughly equivalent to PBO. In May 2008, the Teijin Aramid group announced a "super-fibers" development program. The Teijin emphasis appears to be on computational chemistry to define a solution to high tenacity without environmental weakness.

The materials science of second generation "super" fibers is complex, requires large investments, and represent significant technical challenges. Research aims to develop artificial spider silk which could be super strong, yet light and flexible. Other research has been done to harness nanotechnology to help create super-strong fibers that could be used in future bulletproof vests.

Textile Wovens and Laminates Research

Finer yarns and lighter woven fabrics have been a key factor in improved ballistic results. The cost of ballistic fiber goes up dramatically as yarn size goes down, so it is unclear how long this trend can continue. The current practical limit of fiber size is 200 denier with most wovens limited at the 400 denier level. Three-dimensional weaving with fibers connecting flat wovens together into a 3D system are being considered for both hard and soft ballistics. Team Engineering

Inc is designing and weaving these multi layer materials. Dyneema DSM has developed higher performance laminates using a new, higher strength fiber designated SB61, and HB51. DSM feels this advanced material provides some improved performance, however the SB61 "soft ballistic" version has been recalled. At the Shot Show in 2008, a unique composite of interlocking steel plates and soft UHWMPE plate was exhibited by TurtleSkin. In combination with more traditional woven fabrics and laminates a number of research efforts are working with ballistic felts. Tex Tech has been working on these materials. Like the 3D weaving, Tex Tech sees the advantage in the 3-axis fiber orientation.

Fibers Used

Ballistic nylon (until the 1970) or Kevlar or Spectra (a competitor for Kevlar) or the polyethylene fiber could be used to manufacture bullet proof vests. The vests of the time were made of ballistic nylon & supplemented by plates of fiber-glass, steel, ceramic, titanium, Doron & composites of ceramic and fiberglass the last being the most effective. One should not forget that there is another known para aramid fiber called Twaron, produced in the Netherlands and is a direct competitor of Kevlar.

Manufacturing Process

a. Making the panel cloth- 1. To make Kevlar cloth, the Kevlar yarns are woven in the simplest pattern, plain or tabby weave. which is merely the over & under pattern of threads the interlace alternatively. 2. Unlike Kevlar, the Spectra used in bulletproof vests is usually not woven. instead the strong polyethelene polymer filaments are spun into fibers that are then laid parallel to each other. Resin is used to coat the fibers, sealing them together to form a sheet of Spectra cloth. Two sheets of cloth are then placed at right angles to one another and again bonded, forming a nonwoven fabric that is next sandwiched between two sheets of polyethelene film. The vest shape can then be cut from the material.

b. Cutting The panels-

C. Sewing the ballistic panels-

d. Finishing the vest- The shells for the panels are sewn together in the same standard industrial sewing machines and standard sewing practices. The panels are then slipped inside the shells and the accessories—such as the straps—are sewn on. The finished bulletproof vest is boxed and shipped to the customer.

Working Principle

When a hand gun bullet strikes body armor, it is caught in a "web" of very strong fibers. These fibers absorb and disperse the impact energy that is transmitted to the bullet proof vest from the bullet causing the bullet to deform, otherwise known as a "mushroom". Additional energy is absorbed by each successive layer or material in bullet proof vests until such time as the bullet has been stopped.

Ceramic plates work by locally shattering where the projectile strikes, and are capable of dispersing the energy of the projectile to the point where the bullet has been stopped. Unfortunately, this means that ceramic plates become progressively less capable of stopping additional bullets, and may be rendered unusable after a certain number of hits have been taken.

Developments in Ceramic Armor

Ceramic materials, materials processing and progress in ceramic penetration mechanics are significant areas of academic and industrial activity. This combined field of ceramics armor research is broad and is perhaps summarized best by The American Ceramics Society. ACerS has run an annual armor conference for a number of years and compiled a proceedings 2004–2007. An area of special activity pertaining to vests is the emerging use of small ceramic components. Large torso sized ceramic plates are complex to manufacture and are subject to cracking in use. Monolithic plates also have limited multi hit capacity as a result of their large impact fracture zone These are the motivations for new types of armor plate. These new designs use two- and three-dimensional arrays of ceramic elements that can be rigid, flexible or semi-flexible. Dragon Skin body armor is one of these systems. European developments in spherical and hexagonal arrays have resulted in products that have some flex and multi hit performance. The manufacture of array type systems with flex, consistent ballistic performance at edges of ceramic elements is an active area of research. In addition advanced ceramic processing techniques arrays require adhesive assembly methods. One novel approach is use of hook and loop fasteners to assemble the ceramic arrays.

Nanomaterials in Ballistics

Currently, there are a number of methods by which nanomaterials are being implemented into body armor production. The first, developed at University of Delaware is based on nanoparticles within the suit that become rigid enough to protect the wearer as soon as a kinetic energy threshold is surpassed. These coatings have been described as shear thickening fluids. These nano-infused fabrics have been licensed by BAE systems, but as of mid-2008, no products have been released based on this technology.

In 2005 an Israeli company, ApNano, developed a material that was always rigid. It was announced that this nanocomposite based on tungsten disulfide nanotubes was able to withstand shocks generated by a steel projectile traveling at velocities of up to 1.5 km/s. The material was also reportedly able to withstand shock pressures generated by other impacts of up to 250 metric tons-force per square centimeter (24.5 gigapascals; 3,550,000 psi). During the tests, the material proved to be so strong that after the impact the samples remained essentially unmarred. Additionally, a study in France tested the material under isostatic pressure and found it to be stable up to at least 350 tf/cm² (34 GPa; 5,000,000 psi).

As of mid-2008, spider silk bulletproof vests and nano-based armors are being developed for potential market release. Both the British and American militaries have expressed interest in a carbon fiber woven from carbon nanotubes that was developed at University of Cambridge and has the potential to be used as body armor. In 2008, large format carbon nanotube sheets began being produced at Nanocomp.

Graphene Composite

In late 2014, researchers began studying and testing graphene as a material for use in body armor.

Graphene is manufactured from carbon and is the thinnest, strongest, and most conductive material on the planet. Taking the form of hexagonally-arranged atoms, its tensile strength is known to be 200 times greater than steel, but studies from Rice University have revealed it is also 10 times better than steel at dissipating energy, an ability that had previously not been thoroughly explored. To test its properties, the University of Massachusetts stacked together graphene sheets only a single carbon atom thick, creating layers ranging in thickness from 10 nanometers to 100 nanometers from 300 layers. Microscopic spherical silica "bullets" were fired at the sheets at speeds of up to 3 km (1.9 mi) per second, almost nine times the speed of sound. Upon impact, the projectiles deformed into a cone shape around the graphene before ultimately breaking through. In the three nanoseconds it held together however, the transferred energy traveled through the material at a speed of 22.2 km (13.8 mi) per second, faster than any other known material. If the impact stress can be spread out over a large enough area that the cone moves out at an appreciable velocity compared with the velocity of the projectile, stress will not be localized under where it hit. Although a wide impact hole opened up, a composite mixture of graphene and other materials could be made to create a new, revolutionary armor solution.

Legality

Australia

In Australia, it is illegal to import body armour without prior authorisation from Australian Customs and Border Protection Service. It is also illegal to possess body armour without authorization in South Australia, Victoria, Northern Territory, ACT, Queensland & New South Wales. In 2009 Tasmania considered passing control legislation as well.

Canada

In all Canadian provinces except for Alberta, British Columbia and Manitoba, it is legal to wear and to purchase body armour such as ballistic vests. Under the laws of these provinces, it is illegal to possess body armour without a license (unless exempted) issued by the provincial government. Nova Scotia has passed similar laws, but they are not yet in force.

According to the Body Armour Control Act of Alberta which came into force on June 15, 2012, any individual in possession of a valid firearms licence under the Firearms Act of Canada can legally purchase, possess and wear body armour.

The Netherlands

The civilian ownership of body armour is unregulated in the Netherlands and body armour in various ballistic grades is sold by a range of different vendors, mainly aimed at providing to security guards and VIP's. The use of body armour while committing a crime is not an additional offense in itself, but may be interpreted as so under different laws such as resisting arrest.

United States

United States law restricts possession of body armor for convicted violent felons. Many U.S. states also have penalties for possession or use of body armor by felons. In other states, such as Kentucky, possession is not prohibited, but probation or parole is denied to a person convicted of committing certain violent crimes while wearing body armor and carrying a deadly weapon. Most states do not have restrictions for non-felons.

Hostage Rescue Team agents

Italy

In Italy, the purchase, ownership and wear of ballistic vests and body armor is not subject to any restriction, except for those ballistic protections that are developed under strict military specifications and/or for main military usage, thus considered by the law as "armament materials" and forbidden to civilians. Furthermore, a number of laws and court rulings during the years have rehearsed the concept of a ballistic vest being mandatory to wear for those individuals who work in the private security sector.

European Union

In European Union import and sale of ballistic vests and body armor are allowed in Europe, except protections that are developed under strict military specifications and/or for main military usage, shield above the level of protection NIJ 4, thus considered by the law as "armament materials" and forbidden to civilians. There are many shops in Europe that sell ballistic vests and body armor, used or new.

Implant (Medicine)

An implant is a medical device manufactured to replace a missing biological structure, support a damaged biological structure, or enhance an existing biological structure. Medical implants are man-made devices, in contrast to a transplant, which is a transplanted biomedical tissue. The surface of implants that contact the body might be made of a biomedical material such as titanium, silicone, or apatite depending on what is the most functional. In some cases implants contain electronics e.g. artificial pacemaker and cochlear implants. Some implants are bioactive, such as subcutaneous drug delivery devices in the form of implantable pills or drug-eluting stents.

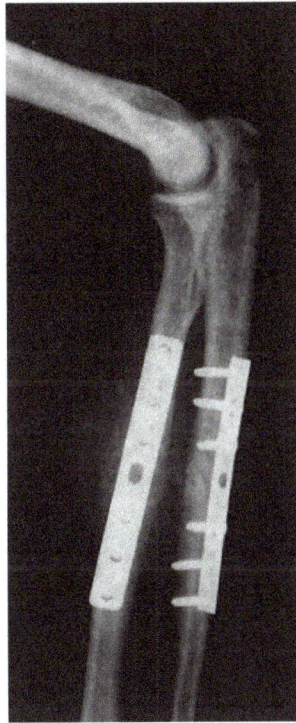

Orthopedic implants to repair fractures to the radius and ulna. Note the visible break in the ulna. (right forearm)

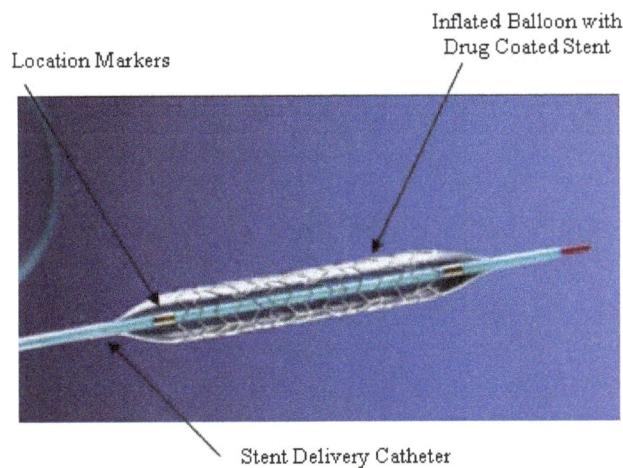

An coronary stent — in this case a drug-eluting stent — is another common item implanted in humans.

Applications

Implants can roughly be categorized into groups by application.

Sensory and Neurological

Sensory and neurological implants are used for disorders affecting the major senses and the brain, as well as other neurological disorders. They are predominately used in the treatment of conditions such as cataract, glaucoma, keratoconus, and other visual impairments; otosclerosis and other hearing loss issues, as well as middle ear diseases such as otitis media; and neurological diseases such as epilepsy, Parkinson's disease, and treatment-resistant depression. Examples include the intraocular lens, intrastromal corneal ring segment, cochlear implant, tympanostomy tube, and neurostimulator.

Cardiovascular

Cardiovascular medical devices are implanted in cases where the heart, its valves, and the rest of the circulatory system is in disorder. They are used to treatment conditions such as heart failure, cardiac arrhythmia, ventricular tachycardia, valvular heart disease, angina pectoris, and atherosclerosis. Examples include the artificial heart, artificial heart valve, implantable cardioverter-defibrillator, cardiac pacemaker, and coronary stent.

Orthopaedic

Orthopaedic implants help alleviate issues with the bones and joints of the body. They're used to treat bone fractures, osteoarthritis, scoliosis, spinal stenosis, and chronic pain. Examples include a wide variety of pins, rods, screws, and plates used to anchor fractured bones while they heal.

Metallic glasses based on magnesium with zinc and calcium addition are tested as the potential metallic biomaterials for biodegradable medical implants·

Contraception

Contraceptive implants are primarily used to prevent unintended pregnancy and treat conditions such as non-pathological forms of menorrhagia. Examples include copper- and hormone-based intrauterine devices.

Cosmetic

Cosmetic implants — often prosthetics — attempt to bring some portion of the body back to an acceptable aesthetic norm. They are used as a follow-up to mastectomy due to breast cancer, for correcting some forms of disfigurement, and modifying aspects of the body (as in buttock augmentation and chin augmentation). Examples include the breast implant, nose prosthesis, ocular prosthesis, and injectable filler.

Other Organs and Systems

Other types of organ dysfunction can occur in the systems of the body, including the gastrointestinal, respiratory, and urological systems. Implants are used in those and other locations to treat conditions such as gastroesophageal reflux disease, gastroparesis, respiratory failure, sleep apnea, urinary and fecal incontinence, and erectile dysfunction. Examples include the LINX, implantable gastric stimulator, diaphragmatic/phrenic nerve stimulator, neurostimulator, surgical mesh, and penile prosthesis.

Classification

United States Classification

Medical devices are classified by the US Food and Drug Administration (FDA) under three different classes depending on the risks the medical device may impose on the user.According to 21CFR

860.3, Class I devices are considered to pose the least amount of risk to the user and require the least amount of control. Class I devices include simple devices such as arm slings and hand-held surgical instruments. Class II devices are considered to need more regulation than Class I devices and are required to undergo specific requirements before FDA approval. Class II devices include X-ray systems and physiological monitors. Class III devices require the most regulatory controls since the device supports or sustains human life or may not be well tested. Class III devices include replacement heart valves and implanted cerebellar stimulators. Many implants typically fall under Class II and Class III devices.

Complications

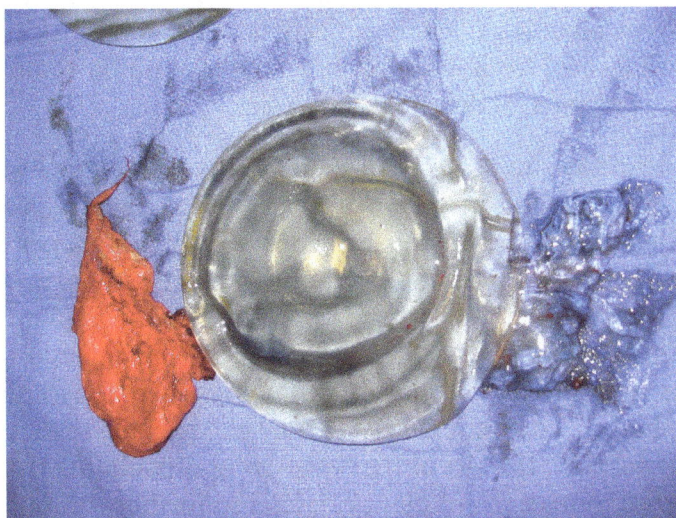

Complications can arise from implant failure. Internal rupturing of a breast implant can lead to bacterial infection, for example.

Under ideal conditions, implants should initiate the desired host response. Ideally, the implant should not cause any undesired reaction from neighboring or distant tissues. However, the interaction between the implant and the tissue surrounding the implant can lead to complications. The process of implantation of medical devices is subjected to the same complications that other invasive medical procedures can have during or after surgery. Common complications include infection, inflammation, and pain. Other complications that can occur include risk of rejection from implant-induced coagulation and allergic foreign body response. Depending on the type of implant, the complications may vary.

When the site of an implant becomes infected during or after surgery, the surrounding tissue becomes infected by microorganisms. Three main categories of infection can occur after operation. Superficial immediate infections are caused by organisms that commonly grow near or on skin. The infection usually occurs at the surgical opening. Deep immediate infection, the second type, occurs immediately after surgery at the site of the implant. Skin-dwelling and airborne bacteria cause deep immediate infection. These bacteria enter the body by attaching to the implant's surface prior to implantation. Though not common, deep immediate infections can also occur from dormant bacteria from previous infections of the tissue at the implantation site that have been activated from being disturbed during the surgery. The last type, late infection, occurs months to years after the implantation of the implant. Late infections are caused by dormant blood-borne bacteria attached to the implant prior to implantation. The blood-borne bacteria colonize on the implant and eventually get released from it. Depending on the type of material used to make the implant, it may be infused with antibiotics to lower the risk of infections during surgery. However, only certain types of materials can be infused with antibiotics, the use of antibiotic-infused implants runs the risk of rejection by the patient since the patient may develop a sensitivity to the antibiotic, and the antibiotic may not work on the bacteria.

Inflammation, a common occurrence after any surgical procedure, is the body's response to tissue damage as a result of trauma, infection, intrusion of foreign materials, or local cell death, or as a part of an immune response. Inflammation starts with the rapid dilation of local capillaries to supply the local tissue with blood. The inflow of blood causes the tissue to become swollen and may cause cell death. The excess blood, or edema, can activate pain receptors at the tissue. The site of the inflammation becomes warm from local disturbances of fluid flow and the increased cellular activity to repair the tissue or remove debris from the site.

Implant-induced coagulation is similar to the coagulation process done within the body to prevent blood loss from damaged blood vessels. However, the coagulation process is triggered from proteins that become attached to the implant surface and lose their shapes. When this occurs, the protein changes conformation and different activation sites become exposed, which may trigger an immune system response where the body attempts to attack the implant to remove the foreign material. The trigger of the immune system response can be accompanied by inflammation. The immune system response may lead to chronic inflammation where the implant is rejected and has to be removed from the body. The immune system may encapsulate the implant as an attempt to remove the foreign material from the site of the tissue by encapsulating the implant in fibrinogen and platelets. The encapsulation of the implant can lead to further complications, since the thick layers of fibrous encapsulation may prevent the implant from performing the desired functions. Bacteria may attack the fibrous encapsulation and become embedded into the fibers. Since the layers of fibers are thick, antibiotics may not be able to reach the bacteria and the bacteria may grow and infect the surrounding tissue. In order to remove the bacteria, the implant would have to be removed. Lastly, the immune system may accept the presence of the implant and repair and remodel the surrounding tissue. Similar responses occur when the body initiates an allergic foreign body response. In the case of an allergic foreign body response, the implant would have to be removed.

Failures

The many examples of implant failure include rupture of silicone breast implants, hip replacement joints, and artificial heart valves, such as the Bjork–Shiley valve, all of which have caused FDA intervention. The consequences of implant failure depend on the nature of the implant and its position in the body. Thus, heart valve failure is likely to threaten the life of the individual, while breast implant or hip joint failure is less likely to be life-threatening.

Artificial Bone

Artificial bone refers to bone-like material created in a laboratory that can be used in bone grafts, to replace human bone that was lost due to severe fractures, disease, etc.

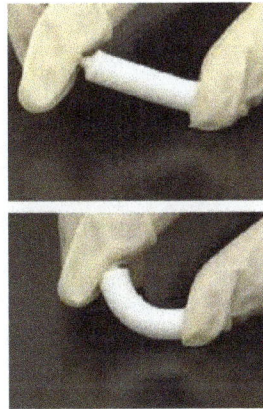

Flexible hydrogel-HA composite, which has a mineral-to-organic matrix ratio approximating that of human bone.

Overview

Bones are rigid organs that serve various functions in the human body (or generally in vertebrates), including mechanical support, protection of soft organs, blood production (from bone marrow), etc. Bone is a very complex tissue: strong, elastic, and self-repairing.

Damaged bone can be replaced with bone from other parts of the body (autografts), from cadavers (allograft), or with various ceramics or metallic alloys. The use of autografts limits how much bone is available.

There has been much research towards creating artificial bone. Richard J. Lagow, at the University of Texas at Austin, developed a way of creating a strong bone-like porous structure from bone powder, which, when introduced in the body, can allow the growth of blood vessels, and which can be gradually replaced by natural bone. Research at the Lawrence Berkeley National Laboratory has resulted in a metal-ceramic composite that has, like bone, a fine microstructure, and which may help create artificial bone. A team of British scientists have developed "injectable bone", a soft substance which hardens in the body. They won the Medical Futures Innovation Award for their discovery, and it is planned to test this material in clinical trials.

Researchers at Columbia University have grown an anatomically correct human jawbone from stem cells, though it was solid bone without the normal accessory tissues such as bone marrow, cartilage, or a connectable blood supply. Other researchers, at the Istec bioceramics laboratory in Italy, have produced a nearly identical substitute for human bone out of rattan wood. The substitute bone has a porous structure permitting blood vessels and other accessory tissues to penetrate it, allowing seamless integration into the host bone. The process has been tested on sheep, who showed no signs of rejection after several months.

Insulator (Electricity)

An electrical insulator is a material whose internal electric charges do not flow freely, and therefore make it nearly impossible to conduct an electric current under the influence of an electric field. This contrasts with other materials, semiconductors and conductors, which conduct electric current more easily. The property that distinguishes an insulator is its resistivity; insulators have higher resistivity than semiconductors or conductors.

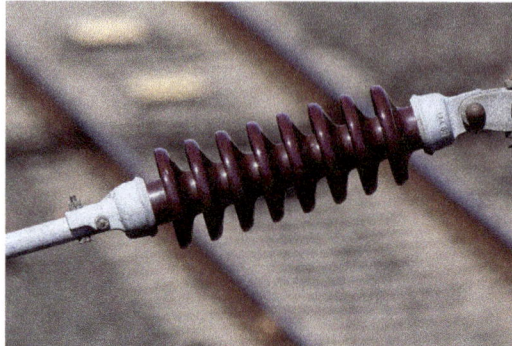

Ceramic insulator used on railways electrified

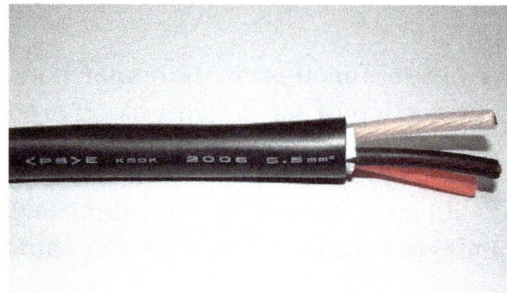

3-core copper wire power cable, each core with individual colour-coded insulating sheaths all contained within an outer protective sheath

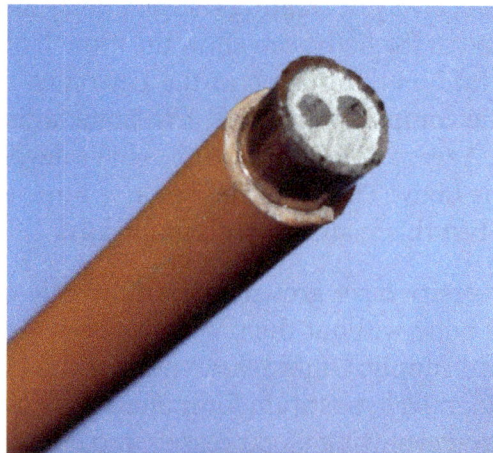

PVC-sheathed Mineral insulated copper cable with 2 conducting cores

A perfect insulator does not exist, because even insulators contain small numbers of mobile charges (charge carriers) which can carry current. In addition, all insulators become electrically conductive when a sufficiently large voltage is applied that the electric field tears electrons away from the atoms. This is known as the breakdown voltage of an insulator. Some materials such as glass, paper and Teflon, which have high resistivity, are very good electrical insulators. A much larger class of materials, even though they may have lower bulk resistivity, are still good enough to prevent significant current from flowing at normally used voltages, and thus are employed as insulation for electrical wiring and cables. Examples include rubber-like polymers and most plastics.

Insulators are used in electrical equipment to support and separate electrical conductors without allowing current through themselves. An insulating material used in bulk to wrap electrical cables or other equipment is called *insulation*. The term *insulator* is also used more specifically to refer to insulating supports used to attach electric power distribution or transmission lines to utility poles and transmission towers. They support the weight of the suspended wires without allowing the current to flow through the tower to ground.

Physics of Conduction in Solids

Electrical insulation is the absence of electrical conduction. Electronic band theory (a branch of physics) says that a charge flows if states are available into which electrons can be excited. This allows electrons to gain energy and thereby move through a conductor such as a metal. If no such states are available, the material is an insulator.

Most (though not all) insulators have a large band gap. This occurs because the "valence" band containing the highest energy electrons is full, and a large energy gap separates this band from the next band above it. There is always some voltage (called the breakdown voltage) that gives electrons enough energy to be excited into this band. Once this voltage is exceeded the material ceases being an insulator, and charge begins to pass through it. However, it is usually accompanied by physical or chemical changes that permanently degrade the material's insulating properties.

Materials that lack electron conduction are insulators if they lack other mobile charges as well. For example, if a liquid or gas contains ions, then the ions can be made to flow as an electric current, and the material is a conductor. Electrolytes and plasmas contain ions and act as conductors whether or not electron flow is involved.

Breakdown

When subjected to a high enough voltage, insulators suffer from the phenomenon of electrical breakdown. When the electric field applied across an insulating substance exceeds in any location the threshold breakdown field for that substance, the insulator suddenly becomes a conductor, causing a large increase in current, an electric arc through the substance. Electrical breakdown occurs when the electric field in the material is strong enough to accelerate free charge carriers (electrons and ions, which are always present at low concentrations) to a high enough velocity to knock electrons from atoms when they strike them, ionizing the atoms. These freed electrons and ions are in turn accelerated and strike other atoms, creating more charge carriers, in a chain reaction. Rapidly the insulator becomes filled with mobile charge carriers, and its resistance drops to a low level. In a solid, the breakdown voltage is proportional to the band gap energy. The air in a region around a high-voltage conductor can break down and ionise without a catastrophic increase in current; this is called "corona discharge". However, if the region of air breakdown extends to another conductor at a different voltage it creates a conductive path between them, and a large current flows through the air, creating an *electric arc*. Even a vacuum can suffer a sort of breakdown, but in this case the breakdown or vacuum arc involves charges ejected from the surface of metal electrodes rather than produced by the vacuum itself. In case of some insulators, the conduction may take place at a very high temperature as then the energy acquired by the valence electrons is sufficient to take them into conduction band.

Uses

A very flexible coating of an insulator is often applied to electric wire and cable, this is called *insulated wire*. Since air is an insulator, in principle no other substance is needed to keep power where it should be. High-voltage power lines commonly use just air, since a solid (e.g. plastic) coating is impractical. However, wires that touch each other produce cross connections, short circuits, and fire hazards. In coaxial cable the center conductor must be supported exactly in the middle of the hollow shield in order to prevent EM wave reflections. Finally, wires that expose voltages higher than 60 V can cause human shock and electrocution hazards. Insulating coatings help to prevent all of these problems.

Some wires have a mechanical covering with no voltage rating—e.g.: service-drop, welding, doorbell, thermostat wire. An insulated wire or cable has a voltage rating and a maximum conductor temperature rating. It may not have an ampacity (current-carrying capacity) rating, since this is dependent upon the surrounding environment (e.g. ambient temperature).

In electronic systems, printed circuit boards are made from epoxy plastic and fibreglass. The nonconductive boards support layers of copper foil conductors. In electronic devices, the tiny and delicate active components are embedded within nonconductive epoxy or phenolic plastics, or within baked glass or ceramic coatings.

In microelectronic components such as transistors and ICs, the silicon material is normally a conductor because of doping, but it can easily be selectively transformed into a good insulator by the application of heat and oxygen. Oxidised silicon is quartz, i.e. silicon dioxide, the primary component of glass.

In high voltage systems containing transformers and capacitors, liquid insulator oil is the typical method used for preventing arcs. The oil replaces air in spaces that must support significant voltage without electrical breakdown. Other high voltage system insulation materials include ceramic or glass wire holders, gas, vacuum, and simply placing wires far enough apart to use air as insulation.

Telegraph and Power Transmission Insulators

Power lines with ceramic insulators in California, USA

Overhead conductors for high-voltage electric power transmission are bare, and are insulated by the surrounding air. Conductors for lower voltages in distribution may have some insulation but are often bare as well. Insulating supports called *insulators* are required at the points where they are supported by utility poles or transmission towers. Insulators are also required where the wire enters buildings or electrical devices, such as transformers or circuit breakers, to insulate the wire from the case. These hollow insulators with a conductor inside them are called bushings.

10 kV ceramic insulator, showing sheds

Material

Insulators used for high-voltage power transmission are made from glass, porcelain or composite polymer materials. Porcelain insulators are made from clay, quartz or alumina and feldspar, and are covered with a smooth glaze to shed water. Insulators made from porcelain rich in alumina are used where high mechanical strength is a criterion. Porcelain has a dielectric strength of about 4–10 kV/mm. Glass has a higher dielectric strength, but it attracts condensation and the thick irregular shapes needed for insulators are difficult to cast without internal strains. Some insulator manufacturers stopped making glass insulators in the late 1960s, switching to ceramic materials.

Recently, some electric utilities have begun converting to polymer composite materials for some types of insulators. These are typically composed of a central rod made of fibre reinforced plastic and an outer weathershed made of silicone rubber or ethylene propylene diene monomer rubber (EPDM). Composite insulators are less costly, lighter in weight, and have excellent hydrophobic capability. This combination makes them ideal for service in polluted areas. However, these materials do not yet have the long-term proven service life of glass and porcelain.

Design

The electrical breakdown of an insulator due to excessive voltage can occur in one of two ways:

- A *puncture arc* is a breakdown and conduction of the material of the insulator, causing an electric arc through the interior of the insulator. The heat resulting from the arc usually damages the insulator irreparably. *Puncture voltage* is the voltage across the insulator (when installed in its normal manner) that causes a puncture arc.

- A *flashover arc* is a breakdown and conduction of the air around or along the surface of the insulator, causing an arc along the outside of the insulator. They are usually designed to withstand this without damage. *Flashover voltage* is the voltage that causes a flash-over arc.

High voltage ceramic bushing during manufacture, before glazing.

Most high voltage insulators are designed with a lower flashover voltage than puncture voltage, so they flash over before they puncture, to avoid damage.

Dirt, pollution, salt, and particularly water on the surface of a high voltage insulator can create a conductive path across it, causing leakage currents and flashovers. The flashover voltage can be reduced by more than 50% when the insulator is wet. High voltage insulators for outdoor use are shaped to maximise the length of the leakage path along the surface from one end to the other, called the creepage length, to minimise these leakage currents. To accomplish this the surface is moulded into a series of corrugations or concentric disc shapes. These usually include one or more *sheds*; downward facing cup-shaped surfaces that act as umbrellas to ensure that the part of the surface leakage path under the 'cup' stays dry in wet weather. Minimum creepage distances are 20–25 mm/kV, but must be increased in high pollution or airborne sea-salt areas.

Cap and pin insulator string (the vertical string of discs) on a 275 kV suspension pylon.

Suspended glass disc insulator unit used in *cap and pin* insulator strings for high voltage transmission lines

Types of Insulators

These are the common classes of insulator:

- *Pin type insulator* - As the name suggests, the pin type insulator is mounted on a pin on the cross-arm on the pole. There is a groove on the upper end of the insulator. The conductor passes through this groove and is tied to the insulator with annealed wire of the same material as the conductor. Pin type insulators are used for transmission and distribution of communications, and electric power at voltages up to 33 kV. Insulators made for operating voltages between 33kV and 69kV tend to be very bulky and have become uneconomical in recent years.

- *Post insulator* - A type of insulator in the 1930s that is more compact than traditional pin-type insulators and which has rapidly replaced many pin-type insulators on lines up to 69kV and in some configurations, can be made for operation at up to 115kV.

- *Suspension insulator* - For voltages greater than 33 kV, it is a usual practice to use suspension type insulators, consisting of a number of glass or porcelain discs connected in series by metal links in the form of a string. The conductor is suspended at the bottom end of this string while the top end is secured to the cross-arm of the tower. The number of disc units used depends on the voltage.

- *Strain insulator* - A *dead end* or *anchor* pole or tower is used where a straight section of line ends, or angles off in another direction. These poles must withstand the lateral (horizontal) tension of the long straight section of wire. In order to support this lateral load, strain insulators are used. For low voltage lines (less than 11 kV), shackle insulators are used as strain insulators. However, for high voltage transmission lines, strings of cap-and-pin (suspension) insulators are used, attached to the crossarm in a horizontal direction. When the tension load in lines is exceedingly high, such as at long river spans, two or more strings are used in parallel.

- *Shackle insulator* - In early days, the shackle insulators were used as strain insulators. But now a day, they are frequently used for low voltage distribution lines. Such insulators can be used either in a horizontal position or in a vertical position. They can be directly fixed to the pole with a bolt or to the cross arm.

- *Bushing* - enables one or several conductors to pass through a partition such as a wall or a tank, and insulates the conductors from it.

- *Line post insulator*

- *Station post insulator*

- *Cut-out*

Cap and Pin Insulators

Higher voltage transmission lines usually use modular *cap and pin* insulator designs *(pictures, left)*. The wires are suspended from a 'string' of identical disc-shaped insulators that attach to each other with metal clevis pin or ball and socket links. The advantage of this design is that insulator strings with different breakdown voltages, for use with different line voltages, can be constructed by using different numbers of the basic units. Also, if one of the insulator units in the string breaks, it can be replaced without discarding the entire string.

Each unit is constructed of a ceramic or glass disc with a metal cap and pin cemented to opposite sides. In order to make defective units obvious, glass units are designed with Class B construction, so that an overvoltage causes a puncture arc through the glass instead of a flashover. The glass is heat-treated so it shatters, making the damaged unit visible. However the mechanical strength of the unit is unchanged, so the insulator string stays together.

Standard disc insulator units are 25 centimetres (9.8 in) in diameter and 15 cm (6 in) long, can support a load of 80-120 kN (18-27 klbf), have a dry flashover voltage of about 72 kV, and are rated at an operating voltage of 10-12 kV. However, the flashover voltage of a string is less than the sum of its component discs, because the electric field is not distributed evenly across the string but is strongest at the disc nearest to the conductor, which flashes over first. Metal *grading rings* are sometimes added around the disc at the high voltage end, to reduce the electric field across that disc and improve flashover voltage.

A recent photo of an open wire telegraph pole route with porcelain insulators. Quidenham, Norfolk, United Kingdom.

In very high voltage lines the insulator may be surrounded by corona rings. These typically consist of toruses of aluminium (most commonly) or copper tubing attached to the line. They are designed to reduce the electric field at the point where the insulator is attached to the line, to prevent corona discharge, which results in power losses.

Typical number of disc insulator units for standard line voltages	
Line voltage (kV)	**Discs**
34.5	3
46	4
69	5
92	7
115	8
138	9
161	11
196	13
230	15
287	19
345	22
360	23

History

The first electrical systems to make use of insulators were telegraph lines; direct attachment of wires to wooden poles was found to give very poor results, especially during damp weather.

The first glass insulators used in large quantities had an unthreaded pinhole. These pieces of glass were positioned on a tapered wooden pin, vertically extending upwards from the pole's crossarm (commonly only two insulators to a pole and maybe one on top of the pole itself). Natural contraction and expansion of the wires tied to these "threadless insulators" resulted in insulators unseating from their pins, requiring manual reseating.

Amongst the first to produce ceramic insulators were companies in the United Kingdom, with Stiff and Doulton using stoneware from the mid-1840s, Joseph Bourne (later renamed Denby) producing them from around 1860 and Bullers from 1868. Utility patent number 48,906 was granted to Louis A. Cauvet on 25 July 1865 for a process to produce insulators with a threaded pinhole: pin-type insulators still have threaded pinholes.

The invention of suspension-type insulators made high-voltage power transmission possible. As transmission line voltages reached and passed 60,000 volts, the insulators required become very large and heavy, with insulators made for a safety margin of 88,000 volts being about the practical limit for manufacturing and installation. Suspension insulators, on the other hand, can be connected into strings as long as required for the line's voltage.

A large variety of telephone, telegraph and power insulators have been made; some people collect them, both for their historic interest and for the aesthetic quality of many insulator designs and finishes. One collectors organisation is the US National Insulator Association, which has over 9,000 members.

Insulation of Antennas

Egg shaped strain insulator

Often a broadcasting radio antenna is built as a mast radiator, which means that the entire mast structure is energised with high voltage and must be insulated from the ground. Steatite mountings are used. They have to withstand not only the voltage of the mast radiator to ground, which can reach values up to 400 kV at some antennas, but also the weight of the mast construction and dynamic forces. Arcing horns and lightning arresters are necessary because lightning strikes to the mast are common.

Guy wires supporting antenna masts usually have strain insulators inserted in the cable run, to keep the high voltages on the antenna from short circuiting to ground or creating a shock hazard. Often guy cables have several insulators, placed to break up the cable into lengths unwanted electrical resonances in the guy. These insulators are usually ceramic and cylindrical or egg-shaped (see picture). This construction has the advantage that the ceramic is under compression rather than tension, so it can withstand greater load, and that if the insulator breaks, the cable ends are still linked.

These insulators also have to be equipped with overvoltage protection equipment. For the dimensions of the guy insulation, static charges on guys have to be considered. At high masts these can be much higher than the voltage caused by the transmitter, requiring guys divided by insulators in multiple sections on the highest masts. In this case, guys which are grounded at the anchor basements via a coil - or if possible, directly - are the better choice.

Feedlines attaching antennas to radio equipment, particularly twin lead type, often must be kept at a distance from metal structures. The insulated supports used for this purpose are called *standoff insulators*.

Insulation in Electrical Apparatus

The most important insulation material is air. A variety of solid, liquid, and gaseous insulators are also used in electrical apparatus. In smaller transformers, generators, and electric motors, insulation on the wire coils consists of up to four thin layers of polymer varnish film. Film insulated magnet wire permits a manufacturer to obtain the maximum number of turns within the available space. Windings that use thicker conductors are often wrapped with supplemental fiberglass insulating tape. Windings may also be impregnated with insulating varnishes to prevent electrical corona and reduce magnetically induced wire vibration. Large power transformer windings are still mostly insulated with paper, wood, varnish, and mineral oil; although these materials have been used for more than 100 years, they still provide a good balance of economy and adequate performance. Busbars and circuit breakers in switchgear may be insulated with glass-reinforced plastic insulation, treated to have low flame spread and to prevent tracking of current across the material.

In older apparatus made up to the early 1970s, boards made of compressed asbestos may be found; while this is an adequate insulator at power frequencies, handling or repairs to asbestos material can release dangerous fibers into the air and must be carried cautiously. Wire insulated with felted asbestos was used in high-temperature and rugged applications from the 1920s. Wire of this type was sold by General Electric under the trade name "Deltabeston."

Live-front switchboards up to the early part of the 20th century were made of slate or marble. Some high voltage equipment is designed to operate within a high pressure insulating gas such as sulfur hexafluoride. Insulation materials that perform well at power and low frequencies may be unsatisfactory at radio frequency, due to heating from excessive dielectric dissipation.

Electrical wires may be insulated with polyethylene, crosslinked polyethylene (either through electron beam processing or chemical crosslinking), PVC, Kapton, rubber-like polymers, oil impregnated paper, Teflon, silicone, or modified ethylene tetrafluoroethylene (ETFE). Larger power cables may use compressed inorganic powder, depending on the application.

Flexible insulating materials such as PVC (polyvinyl chloride) are used to insulate the circuit and prevent human contact with a 'live' wire – one having voltage of 600 volts or less. Alternative materials are likely to become increasingly used due to EU safety and environmental legislation making PVC less economic.

Class 1 and Class 2 Insulation

All portable or hand-held electrical devices are insulated to protect their user from harmful shock.

Class 1 insulation requires that the metal body and other exposed metal parts of the device be connected to earth via a *grounding wire* that is earthed at the main service panel—but only needs basic insulation on the conductors. This equipment needs an extra pin on the power plug for the grounding connection.

Class 2 insulation means that the device is *double insulated*. This is used on some appliances such as electric shavers, hair dryers and portable power tools. Double insulation requires that the devices have both basic and supplementary insulation, each of which is sufficient to prevent electric shock. All internal electrically energized components are totally enclosed within an insulated body that prevents any contact with "live" parts. In the EU, double insulated appliances all are marked with a symbol of two squares, one inside the other.

Airframe

Airframe diagram for an AgustaWestland AW101 helicopter

The airframe of an aircraft is its mechanical structure. It is typically considered to include fuselage, wings and undercarriage and exclude the propulsion system. Airframe design is a field of aerospace engineering that combines aerodynamics, materials technology and manufacturing methods to achieve balances of performance, reliability and cost.

History

Wellington Mark X showing the geodesic airframe construction and the level of punishment it could withstand while maintaining airworthiness

Modern airframe history began in the United States when a 1903 wood biplane made by Orville and Wilbur Wright showed the potential of fixed-wing designs. Many early developments were spurred by military needs during World War I. Well known aircraft from that era include the Dutch designer Anthony Fokker's combat aircraft for the German Empire's *Luftstreitkräfte*, and U.S. Curtiss flying boats and the German/Austrian Taube monoplanes. These used hybrid wood and metal structures. During the war, German engineer Hugo Junkers pioneered practical all-metal airframes as early as late 1915 with the Junkers J 1 and developed further with lighter weight duralumin in the airframe of the Junkers D.I of 1918, whose techniques were adopted almost unchanged after the war by both American engineer William Bushnell Stout and Soviet aerospace engineer Andrei Tupolev. Commercial airframe development during the 1920s and 1930s focused on monoplane designs using radial piston engines. Many, such as the Ryan model flown across the Atlantic by Charles Lindbergh in 1927, were produced as single copies or in small quantity. William Stout's designs for the all-metal Ford 4-AT and 5-AT trimotors, Andrei Tupolev's designs in Joseph Stalin's Soviet Union for a series of all-metal aircraft of steadily increasing size, culminating in the enormous, eight-engined *Maksim Gorky* (the largest aircraft of its era), and with Donald Douglas' firm's development of the iconic Douglas DC-3 twin-engined airliner, were among the most successful designs to emerge from the era through the use of all-metal airframes. The original Junkers corrugated duralumin-covered airframe philosophy culminated in the 1932-origin Junkers Ju 52 trimotor airliner, used throughout World War II by the Nazi German Luftwaffe for transport and paratroop needs.

During World War II, military needs again dominated airframe designs. Among the best known were the US Douglas C-47, Boeing B-17, North American B-25 and Lockheed P-38, and British Vickers Wellington that used a geodesic construction method, and Avro Lancaster, all revamps of original designs from the 1930s. The wooden composite construction high performance fighter-bomber de Havilland Mosquito was developed during the war. The first jets were produced during the war but not made in large quantity. The Boeing B-29 was designed to be a high altitude bomber, the first with a pressurised fuselage.

Postwar commercial airframe design focused on larger capacities, on turboprop engines, and then on jet (turbojet, later turbofan) engines. The generally higher speeds and stresses of turboprops and jets were major challenges. Newly developed aluminum alloys with copper, magnesium and zinc were critical to these designs. The Lockheed L-188 turboprop, first flown in 1957, used some of these materials and became a costly lesson in controlling vibration and planning around metal fatigue.

DH106 Comet 3 G-ANLO demonstrating at the 1954 Farnborough Airshow

The de Havilland Comet was the world's first commercial jet airliner to reach production. It first flew in 1949 and was considered a landmark in British aeronautical design. After introduction into commercial service, early Comet models suffered from catastrophic airframe metal fatigue, causing a string of well-publicised accidents. The Royal Aircraft Establishment investigation at Farnborough, founded the science of aircraft crash reconstruction. Over 3000 cycles of pressurisation later, in a specially constructed pressure chamber, airframe failure was found to be due to stress concentration, a consequence of the square shaped windows. The windows had been engineered to be glued and riveted, but had been punch riveted only. Unlike drill riveting, the imperfect nature of the hole created by punch riveting may cause the start of fatigue cracks around the rivet.

Eventually Boeing in the U.S. and Airbus in Europe became the dominant assemblers of large airframes, known as wide-body aircraft. Numerous manufacturers in Europe, North America and South America took over markets for airframes designed to carry 100 or fewer passengers. Many manufacturers produce airframe components.

Present and Future

Rough interior of a Boeing 747 airframe

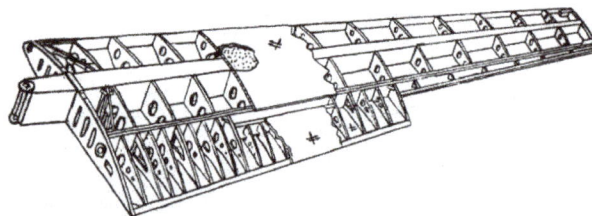

Wing structure with ribs and one spar

Four major eras in commercial airframe production stand out: all-aluminum structures beginning in the 1920s and directly inspired by Hugo Junkers all-metal designs from as far back as 1915, high-strength alloys and high-speed airfoils beginning in the 1940s, long-range designs and improved efficiencies beginning in the 1960s, and composite material construction beginning in the 1980s, partly pioneered by Burt Rutan's designs. In the latest era, Boeing has claimed a lead, designing its new 787 jetliner with a one-piece carbon-fiber fuselage, said to replace "1,200 sheets of aluminum and 40,000 rivets." The Airbus A380 is also built with a large proportion of composite material.

Cirrus Aircraft's SR20 design, certified in 1998, was the first general aviation aircraft manufactured with all-composite construction, followed by several other light aircraft in the 2000s.

Airframe production has become an exacting process. Manufacturers operate under strict quality control and government regulations. Departures from established standards become objects of major concern. The crash on takeoff of an Airbus A300 in 2001, after its tail assembly broke away from the fuselage, called attention to operation, maintenance and design issues involving composite materials that are used in many recent airframes. The A300 had experienced other structural problems but none of this magnitude. The incident bears comparison with the 1959 Lockheed L-188 crash in showing difficulties that the airframe industry and its airline customers can experience when adopting new technology.

References

- Williams, Allan (2003). The Knight and the Blast Furnace: A History of the Metallurgy of Armour in the Middle Ages & the Early Modern Period. Boston: Brill Academic Publishers. ISBN 978-90-04-12498-1.

- Metz, Leon Claire (2003). The Encyclopedia of Lawmen, Outlaws, and Gunfighters. Checkmark Books. pp. 172–173. ISBN 0-8160-4543-7.

- Erwin, Richard E. (1993). The Truth about Wyatt Earp (2nd ed.). Carpinteria, CA: O.K. Press. ISBN 9780963393029.

- Hollington, Kris (2008). Wolves, Jackals, and Foxes: The Assassins Who Changed History. St. Martin's Press. ISBN 9781429986809.

- Wong, J.Y.; Bronzino, J.D.; Peterson, D.R., eds. (2012). Biomaterials: Principles and Practices. Boca Raton, FL: CRC Press. p. 281. ISBN 9781439872512. Retrieved 12 March 2016.

- McLatchie, G.; Borley, N.; Chikwe, J., eds. (2013). Oxford Handbook of Clinical Surgery. Oxford, UK: OUP Oxford. p. 794. ISBN 9780199699476. Retrieved 12 March 2016.

- Duke, J.; Barhan, S. (2007). "Chapter 27: Modern Concepts in Intrauterine Devices". In Falcone, T.; Hurd, W. Clinical Reproductive Medicine and Surgery. Elsevier Health Sciences. pp. 405–416. ISBN 9780323076593. Retrieved 12 March 2016.

- Black, J. (2006). Biological Performance of Materials: Fundamentals of Biocompatibility. Boca Raton, FL: CRC Press. p. 520. ISBN 9780849339592. Retrieved 12 March 2016.

- Dee, K.C.; Puleo, D.A.; Bizios, R. (2002). An Introduction to Tissue-Biomaterial Interactions. Hoboken, NJ: Wiley-Liss. p. 248. ISBN 9780471461128. Retrieved 12 March 2016.

- Syring, G. (6 May 2003). "Overview: FDA Regulation of Medical Devices". Quality and Regulatory Associates, LLC. Retrieved 12 March 2016.

- "Classify Your Medical Device". FDA.gov/MedicalDevices. Food and Drug Administration. 29 July 2014. Retrieved 12 March 2016.

- "Tests prove that a bulletproof silk vest could have stopped the first world war". The Guardian. July 29, 2014. Retrieved November 13, 2015.

- "Could the Bulletproof Silk Vest Stop the Assassination Which Sparked WWI? Tests Say Yes". War History Online. August 8, 2014. Retrieved November 13, 2015.

- "Turning wood into bone". BBC News. 8 January 2010. Archived from the original on 9 January 2010. Retrieved 8 January 2010.

- Florence Graves and Sara K. Goo (Apr 17, 2006). "Boeing Parts and Rules Bent, Whistle-Blowers Say". Washington Post. Retrieved April 23, 2010.

Permissions

Index

A

Amber, 139

Amorphous, 1, 3, 5, 15, 19, 22, 25, 60, 64, 70, 85-86, 90-93, 99-100, 114-117, 121, 166, 169-170, 195, 199, 203

Audio Signal, 142

B

Bakeware, 167

Ball Bearings, 13, 17, 72

Ballistic Protection, 12, 26, 111, 232-233, 243, 245

Batching, 19-20

Beryllia, 12

Binder, 20, 37, 40, 50, 199, 206

Biocompatibility, 166, 168, 269

Bone China, 12

Boride, 12

Boron Carbide, 13, 17, 63, 66-68, 232, 241

Brake Disk, 95

Brittle Materials, 4, 163, 199

Building Construction, 181-182, 206

C

Carbon Fiber, 1, 123, 204, 209-210, 214-215, 248

Cartesian Tensor, 143

Ceramic Art, 3

Ceramic Shell Casting, 36

Ceramography, 4, 29-32

Clay, 2-3, 12, 14-15, 26, 35, 50, 74, 78-80, 180, 182-185, 187-188, 199, 206, 236-237, 239-240, 242-243, 259

Coil Construction, 36

Composite Material, 4, 6, 10, 200, 205-206, 208-209, 268

Cooktop, 167-168

Crack Propagation, 22, 33, 88, 108, 189, 212, 222

Crystalline Ceramics, 3, 97, 99, 103, 108

Cubic, 15, 23, 31-32, 55, 63-66, 68, 71, 96, 108-109, 124, 144, 206

D

Dielectric, 4, 8, 33-34, 65, 74-76, 142, 144-145, 154, 166, 259, 265

E

Earthenware, 10, 12, 180, 182-183

Elasticity, 1, 4-5, 141, 159-161, 197, 221, 234

Elastomer, 122

Erosion, 3, 102, 215

F

Ferroelectric, 7, 75-76, 144, 146-147, 220

Ferroelectrics, 141

Fiberglass, 1, 206, 210, 212, 214, 231, 247, 265

Forming, 3, 6, 11, 13, 18-21, 25, 35-37, 39, 41, 43, 45, 47, 49, 51, 53, 55, 57, 59, 61, 66, 83, 86-87, 98, 103, 108-109, 117, 121-123, 129, 166, 169, 173-176, 189, 213, 215, 247

Fractography, 5

Fracture Toughness, 4, 6, 9, 13, 21, 28-29, 64, 68, 73, 84, 87, 89, 92, 106, 208, 241

G

Glass-ceramics, 18-19, 21, 98, 100, 108, 166-168, 171, 222

Grain Boundary Strengthening, 22, 112-113, 116, 120

H

Hermetic Sealing, 166

I

Incandescent, 53, 65, 172, 177

Induction Stove, 168

Infrared, 8, 16, 27, 34, 65, 97-98, 101-102, 104-110, 152, 167-168, 179

Ionizing Radiation, 68

Ir Night Vision, 16, 97

L

Leds, 6, 8, 202

M

Mechanical Stress, 76, 94, 139, 142, 144

Metastable, 11, 24, 57-58, 64, 121

Microstructure, 4, 21-22, 28-29, 32-33, 41-42, 44, 55-57, 61-62, 77, 97, 111, 115, 193-195, 197, 255

Milling, 19, 25, 73, 85

Missile Guidance, 16, 97, 105

Mixing, 19-20, 25, 47, 50, 74, 109, 122, 130, 213

N

Nozzle, 73, 139

P

Paraffin Wax, 36

Piezoelectricity, 7, 15, 126, 139-140, 142, 144-148, 151 152, 164

Pigmentation, 166

Pinching, 36

Plasticity, 4-5, 50, 113, 121, 123, 125

Polyvinyl Alcohol, 21

Porosity, 3-4, 19, 21-23, 25, 32-33, 38, 42, 47-53, 55, 57, 84, 86-88, 90-92, 95, 100, 102, 107, 166, 170, 222

Pottery, 2, 10-11, 14, 21, 35, 50, 182-184, 188

R

Radiant Heating, 168

S

Self-assembly, 23-24, 29

Semiconductor, 11, 53, 68, 70, 90, 109-111, 147, 172, 202, 204, 219

Sherds, 14

Silicide, 12, 175-176

Silicon Carbide, 3, 5, 10-11, 15, 17, 24, 26, 30, 84, 86, 90, 92-94, 199-200, 202, 209, 212, 232, 241

Silicon Nitride, 2, 7, 11, 17, 30, 63, 69-75, 84, 86

Sintering, 3, 11, 15, 19-23, 25, 35-36, 46-56, 60-61, 70, 75 77, 85-87, 91, 95, 100-103, 106-107, 109-111, 130-131, 166-167, 189, 199, 220, 222

Slip Casting, 4, 20, 35, 54, 108-109

Solid State Chemistry, 4

Space Shuttle Program, 12, 26

Spark Plug, 20

Strain Hardening, 162

Stress Concentration, 64, 268

Superconductivity, 6-7, 126-129, 131-138, 164, 166, 204, 218

T

Technical Ceramics, 12, 26, 36, 83-84, 92

Tensile Strength, 4-5, 85, 88-90, 117, 122, 155, 161, 177 178, 202, 232, 237, 241, 249

Terracotta, 180-188, 190

Tetragonal, 24, 55, 63-64, 124, 129-130, 134, 137, 143, 146

Tetragonal Zirconia Polycrystal, 64

Thermal Barrier Coating, 65

Thermal Conductivity, 25, 41, 49, 65, 91, 99, 106, 127, 171, 189

Tissue Engineering, 13, 18, 168-169, 198, 219

Topaz, 140, 145

Translucent, 8, 100, 167

Transparent Armour, 16

Tungsten Carbide, 3, 15, 25, 36, 51, 72-73, 199, 241

U

Uniaxial Tensile Testing, 161

V

Van Der Waals Forces, 22, 222

Voigt Notation, 143

Z

Zirconia, 11-12, 15, 24-26, 30, 32-33, 35-36, 50, 54-55, 63 66, 84-85

www.ingramcontent.com/pod-product-compliance
Lightning Source LLC
Chambersburg PA
CBHW061313190326
41458CB00011B/3795